California Contractor's **Desktop General Counsel**

California Contractor's DESKTOP GENERAL COUNSEL

What You Need To Know About California Construction Law

John P. McGill

M3
Publisher

California Contractor's
DESKTOP GENERAL COUNSEL
What You Need To Know About California Construction Law

For information, or to order additional copies, please contact:
M3-PUBLISHER
1436 Second St #207
Napa CA 94558
www.M3-Publisher.com
or call 707 337 1932

Edited by Walter Klein (www.kleineedit.com)
Cover design by KM Design- Boston MA. and Frye's Printing Napa CA
Interior design by Cypress House, Fort Bragg CA.
Indexing by Mary Harper at Access Point Indexing

ISBN: 978-0-9840191-3-7
Library of Congress Control Number: 2011961983

Printed in the United States of America

2 4 6 8 9 7 5 3

First edition

For

Wanda, Kaelin & RyLee

The Only Three Reasons I Do Anything

And Not Necessarily In That Order

Acknowledgements

THIS BOOK WAS A LONG TIME COMING and is the result of a lot of assistance- intended and unintended. Once it was written and nearly complete, I had the good fortune to find Walter Kleine to do the editing on it; an excellent choice. The interior type design, which I had the audacity to believe I could do myself (not likely!), was done by Cypress House, a further stroke of good fortune. The cover design is my idea, carried out to exactly what I really wanted despite my inability to articulate it clearly, by KM Design in Boston and Elizabeth at Frye's Printing in Napa CA; thanks for your help and patience.

Special thanks to Bryan Richmond at Sanders Jacobs Insurance in Napa who gave me some excellent information and assistance with the section on insurance. Like the book says, find a good agent to use- I did. And to the folks at Arntz Builders Inc. who took me on as general counsel and then taught me the *in's and out's* of public works contracting- thanks very much for putting up with me.

The other major thanks goes to all of the contractors and building professionals, and attorneys too, that I have had the good fortune to work with over the many years. Again, thank you one and all. I am a better attorney and a better craftsman for all of it.

"Knowledge is of two kinds. We know a subject ourselves, or we know where we can find information upon it."

— Samuel Johnson

✳

"The difficult task, after one learns how to think like a lawyer, is relearning how to write like a human being."

— Unknown

✳

"You can starve one attorney, but you can't starve two."

— John McGill

Contents

California Contractor's

DESKTOP GENERAL COUNSEL

Foreword

When I first had the idea for this book, I thought it would likely be unnecessary. Certainly, I thought, there were already lots of books out there for contractors that explained the law and how it applied to the construction industry in California. Strangely there weren't; and what books there were, were written by attorneys for attorneys. Even those that were supposedly for contractors were not especially useful in any practical way.

As in-house general counsel for a large public works general contractor, I was asked/confronted with legal issues that I needed to explain in terms that the decision-makers could make sense of. Telling them what a statute said or quoting the statute verbatim was not what they wanted or expected. I needed to explain how and why a particular rule would impact on what we wanted to do or were intending to do. These were smart businessmen who asked difficult questions and didn't want to listen to a bunch of legalese. They needed to understand the rule, not have it read back to them so they could decipher it. Once they had the information they could make the decision; my job was to give them that information. Needless to say I also learned a lot.

And that's what this book is about: explaining the law so that an informed business decision can be made. Some of what is included is fairly basic; other things, not so much. Like anything else, we sometimes take the fundamentals for granted. But the funny thing is that most of the mistakes that are made by contractors, and that are very costly mistakes indeed, are mistakes on very basic things. So the book covers the basics and the more esoteric; what you know and what you need to know.

The book is divided into five major topics: licensing because that is a critical prerequisite; contracts because understanding contracts, and what they are, is essential if that is how you are going to make your living;

public works contracting, a special kind of contracting to be sure; private works contracting, which is the most common form of the business; and litigation because, as they say, s*** happens.

These topics are broken down into sections and the sections follow in what should be a logical order and sequence. Scattered throughout the sections are NOTES that offer suggestions and warnings based on real situations that can and do occur; your basic *head's up*.

After nearly 40 years in the industry, first as a carpenter, then a contractor, and now as an attorney representing contractors, I still think the best way to spend a day is walking a jobsite and watching things get built. Contractors and construction workers are some of the best business people and most creative types there are. There is also nothing more individually rewarding than driving by a building and knowing that you built it. Construction is hard work, but it's also a lot of fun.

This book is for contractors and for anyone working in the industry. It contains no infomercials for any firm or company. It is pure and simple legal information for the construction professional. I hope it ends up on your desk, is well thumbed and referred to often, and eventually comes to resemble the tools used on your projects- worn and dirty, but still able to get the job done.

JPM 2012

Introduction

THERE'S ONE THING THAT we do better in this country than most, if not all, of the rest of the world. We build. We build roads and highways, dams and bridges, houses and skyscrapers, train stations and boat docks, tree houses and doghouses. We can build anything, and we have, and what we build, we build really well.

There's another thing, too, that we do in this country, probably as much or more than any other country; we make laws. We have a law for nearly everything, and if we don't have a law, well, there *oughta* be a law about that too. And there likely is, or soon will be.

In the construction industry, the laws are well-intentioned but sometimes not very well thought out. Often one group is protected against another's perceived wiles and shiftiness, and so the thumb is on the scale from the beginning. The supposed weaker party is not so much protected as given the opportunity to do its own mischief.

The laws for construction are made by people who, for the most part, don't have a clue about what it is that a contractor does or what construction is about, and even worse, those laws are often enforced by people that don't know why the law is there in the first place or what it is protecting against. All this is well-intended, but it can be a bit of an *Alice Through The Looking Glass* experience the first time you have to deal with it. Nothing seems to be what you expect after you fall down the rabbit hole of the law.

For one thing, contractors who find themselves in a lawsuit are typically represented by someone who doesn't understand what the contractor does, who speaks a language that sounds like English but isn't, and these people seem unable to complete any task without causing at least another two or three more to be required. Contractors make their money by being efficient and keeping to schedules; attorneys bill in increments of 1/10th of an hour. There is a serious disconnect here, and it is going to cost the contractor a lot of money.

If you are unlucky enough to end up in a trial, the rightness or wrongness of your position is irrelevant, and the Red Queen may very well have the verdict before the first witness is even called. Your case is going to be determined by someone who may have even less understanding of the construction industry than your attorney, but who will nonetheless be guided by clever displays of semantics and logical gymnastics that are actually the rules of evidence, and they will use these rules to decide your case.

The judge or jury, euphemistically called the trier of fact, will, after this complete disruption of communication, be educated by a careful orchestration of facts that may or may not be accurate or relevant, and that are presented in a way that will sound more like theater than an effort to ferret out the truth. Then, with an astonishing agility that will leave either you or your opponent, or both, in complete awe, and by incredible but quite agile leaps of faith and logic, the trier of fact will make a decision, a verdict will be submitted, and the case will be over. Sometimes the decision is correct, sometimes not, but it is always the end of the case.

Until the appeal is taken that is; but that's another story and a very expensive one that you don't want to hear. It's a quite remarkable system that we have, and even more astounding is the fact that it works — for the most part. It really is a crapshoot whether the right result is reached at the trial court level. The appellate courts usually get it right, and that's why we have the appellate court; to correct the trial court's errors. Then again, appellate court judges were trial court judges once, so sometimes even that court gets it wrong.

The moral of the story, then, is to stay out of court. Remember, you can starve one attorney, but you can't starve two.

So, how do you stay out of court? Luck mainly, and a lot of forward planning. You need to be prepared, and by being prepared you can usually convince the other side that, at best, they will come away with a Pyrrhic victory; one where the treasure gained is not worth the treasure spent.

So what is all of this leading to? Why write another book on construction law? What's the point?

First, this book is not intended to make you an attorney. If you are an attorney, this book is not intended for you. Any contractor that sees an attorney reading this book should very gently, but persuasively, insist that they put it down and step away slowly. On the other hand, if you see a contractor reading this book and then hear them talking like an attorney (i.e. big words used in sentences that don't follow any recognizable rules of grammar) ignore them or just laugh loudly in their direction and walk away. Shake your head for dramatic effect, and to emphasize your displeasure.

Seriously, this book is intended to be a reference for contractors that will explain the basics of construction law, the legal terms and concepts, and how they apply to you, the contractor. If you have a basic understanding of how the law works and how the courts have applied the rules, and what the rules are in the first place, you are better able to anticipate legal pitfalls and either avoid them or mitigate the problems when they become unavoidable. And, if you are in a situation where a gaggle of lawyers are blathering away, you will at least be able to understand what the heck they are talking about.

The intent of this book is to provide you with a basic reference, using suggestions, explanations, and advice that can (hopefully) steer your company away from legal problems, or at least get you prepared to navigate through them until you get an attorney on board. And of course, you will be able to speak *pidgin* legalese that will help when you speak to your attorney. Remember, attorneys are on a different clock then you are, so the faster you get your point across the more cash you will have left in your wallet.

Again though, you will not become a lawyer by reading this book. You should not try to practice law after reading this book, except in the privacy of your own home. What should result from this book is an ability to hold an intelligent conversation with an attorney, assuming the attorney has a clue about construction, and what contractors do in the first place. Many attorneys do, and many are very good, very smart, and very nice people. Just as many (and maybe more) are not, so the more you know about the basics of the law, the better you can judge who you may want representing you, if that ever becomes the situation. With any luck, it will never be the case.

There are a lot of other reference books on construction law that are available. Depending on how much more information you want, and how in-depth you want to get, there are any number of books that you will find to slake your thirst for the law. Keep in mind though, that sometimes the books available, even those that are ostensibly directed at contractors, are fairly technical. Attorneys use them as much, and likely more, than contractors. It all depends on how much information you need or want. If the question you need an answer to is complicated, then the better approach is to spend your money on a good lawyer who can give you the answer that you need, even if it may not be the one that you want, rather than trying to find the answer by reading a bunch of reference material.

This book is intended as a contractor's desk reference, hence the name *Desktop General Counsel*. It is, hopefully, an easy read, an easy reference, and a convenient size to keep handy for a quick answer or a suggested approach. It does not have to be updated every year because much of what is included is the law as it has been for years, and the rest is common sense applied to circumstances that might arise on a typical project.

Aspects of the law may change, and these changes may be important to know about, but they should not require a bunch of updates. There may be another edition eventually, but not every year. Much of the law is, as we attorneys like to say, well settled. We intend to post reviews of new cases on the website at www.M3-publisher.com. Visit it to check for any recent changes and for new cases that might have an impact on the industry. The cases themselves are sometimes fun to read, especially in summary form, so look them over if you have a few minutes. Don't waste time on the full cases, though. Attorneys read these. You're a contractor; go build something.

Keep in mind, too, that even if there is a decision that appears to change the way the law is applied, unless it is from the California Supreme Court or the US Supreme Court, it may not be that conclusive on the issue. Even if the state or federal Supreme Court weighs in on the issue, if the State

Legislature or Congress disagrees, they can change the law. The point is, nothing is ever really final; well, one thing is, but we won't go there.

The Contractors State License Board website is also a great source of information. (www.cslb.ca.gov) Check it out, get familiar with what is on it, put it on your *Favorites* and bring it up once in awhile, just to make sure you know of any big changes. Contractors need to stay current on licensing law, so check the CSLB website. The website also has a lot of the forms that you might need in the course of your business, and these can be either downloaded or printed out. Use them.

In short, it is important that you know the basics and also keep abreast of the changes in the law and the industry. Forewarned is forearmed, as they say.

Now, down the rabbit hole we go...

Chapter 1

Licensing

Who Needs a License?

IF YOU ARE DOING any work by agreement for someone else, and the cost of that work is $500 or more, you need a license. No exceptions, no excuses, no explanations. It doesn't matter if you write out five separate agreements, and agree to be paid $100 for each, or if you have one contract for the $500. You need a license, and very bad things can happen to you if you don't have one. Don't play games with this; get a license.

Yes, it's an inconvenience and a pain, and yes, there is a test, and yes, it is kind of pricey to take the course or buy the books to get ready for the test, but if you intend to make a living in contracting, get the license and spend the money to do it.

What kind of license you get is important, too. There are four types of licenses

A – Engineering and Grading
B – General Contractor
C – Specialty Contractor
D – Super specialty

If you have a backhoe and dump truck, and you move dirt for a living, you probably want an A license.

If you build or run the type of projects that have a lot of different trades going on, at a minimum two different ones not including framing, then you probably want a B license.

If you just do painting, plumbing, electrical, or any specialized trade activity, you want a C license.

Finally, if you're really, really, really specialized, you want a C-61 specialty license that lets you get any of a number of sub-category licenses. These are D licenses for the singular types of things that some folks just can't get enough of.

So how does all this work? Can you do things that are not in the category that you are qualified (licensed) in? The answer is a definite maybe. For example, an A license can't be a General Contractor. But a B license General Contractor can move dirt, if it's related to the work that the General is doing under the B license and provided the B license holder's principal work is in constructing buildings and not just moving dirt.

For example, assume that the General is building a house and he wants to take his personal D-9 to the job and clear the residential lot, just for kicks, because he likes pushing dirt around. He can do it even if he only has the B license. That same General can also take his trusty D-9 and do a site development even if he never puts up a single building. This is only the case though if the B license holder's primary work is to put up buildings of one sort or another. The license law requires that the B license holder's principal work be in constructing buildings. If the only work this fellow does is site work, then he is not properly licensed and needs an A license.

Likewise, the C license specialty contractor can sometimes do work outside of her trade if it's "incidental" to the work that is being done under the C license. Let's say our C license holder is a tile setter (C-54) doing a bathroom remodel. She is hired to move a tub, replace it with a shower stall, and then tile the floor and do a bit of wainscot around the walls of the bathroom. The problem is that the tub's plumbing is in exactly the wrong place for the new shower. It has to be moved. Can she move it? Probably yes because it is incidental to the tile work, and assuming she knows what and how to do the plumbing, she can do this work.

What she cannot do is rip out the walls, re-route the sink and shower plumbing, put in an 'instant hot,' add another sink, relocate the toilet, add a closet, put up the gypboard, paint the room, install the cabinets, add new electrical, put in the shower glass, and then tile the shower, floor, and wainscot. She can do the last three, but the rest of it is not incidental and she needs a B license, or at least a whole stash of C specialty licenses.

More likely, she needs to sub it out to other contractors, which would also require that she have a B license, or, even more likely, the owner needs to hire a bunch of other contractors to assist our tile setter.

In a similar but even more limited way, the C license can do the work of a D specialized license, but the D license can't do all that the C license holder can do. For example, a painter (C-33) can prep the surfaces by sandblasting them. With the C license, the painter can then paint that surface. However, the D-38 sandblast specialty license can sandblast the surface, but the D license holder cannot paint the surface. Painting is not within the D license category.

In short, the more specialized your license is, the less you can do. What you do is likely done very well, but if you hold only a D license, then whatever the License Board category says you can do, is all that you can do. If you are C license, you can do more. If you are a B license, and the job requires two or more unrelated trades, you can take it as a General. However, if it's just two trades — plumbing and framing — you can't take it as a General. You can do it as a framer, or you can do it as a plumber, but only if you have a B license for the framing and a C-36 for the plumbing. If you have both licenses, go for it. If you only have an A license, forget it — no bottom dumps allowed.

The best place to find out what work a particular license can do is through the Contractors State License Board. Use the website, www.cslb.ca.gov, to investigate whatever questions you have. If you need more information or a better explanation, you can write to the Board and they will respond. It may take a little time for them to get back to you, but they will. You can also call and ask for a fax number to one of the license classification specialists and fax them directly. The people who work at the Board are generally very helpful and pretty nice, but they are busy. If you don't get a response, call again. Be a pest; they don't seem to mind and they will give you a written response, which is really what you want. Don't just ask the person answering the phone and take their word for it; get it in writing.

Who Is "You"?

So you know you need a license, but who is "you?" You can be an individual, a partnership, a corporation, or a joint venture. You can also be an LLC, that happened as of 2012.

If you are a sole proprietor, then you need a license. If you are a partnership, then one or more of the partners needs to have a license or licenses for the type of work the partnership wants to do. If you are a corporation, you need qualifiers — a Responsible Managing Officer (RMO) or a Responsible Managing Employee (RME) — someone who has passed the test for the particular license category for the work your company wants to do. A corporation can have a number of people that are qualifiers for any number of different categories. One person can hold the A license, another the B license, and another a C license, or one person can hold all three if they take and pass the tests.

If the qualifier is not an officer of the company, then that person can qualify for the license as a Responsible Managing Employee (RME). The RME needs to devote a majority of his/her time managing the job where the particular license is being used. Don't think that just because Jethro has a drywall license that you can use his license but send him off to some backwater project and never have to deal with him. The RME has to be involved in the work, or you may not be in compliance with the licensing law.

> **NOTE:** There are some fairly technical rules that apply if the qualifier will be involved qualifying more than one company. For example, in order to qualify for another company, the other company must have common ownership of at least 20% of the equity shares. This is (presumably) so that the qualifier him/herself is not simply starting up a bunch of corporations with only themselves as the shareholder. Likewise, if the qualifier is not a proprietor, general partner, or joint licensee, then a $12,500 bond needs to be posted in addition to the contractor's license bond. This bond requirement is waived if the qualifier owns at least 10% of the voting stock of

the company or if the individual qualifier has personally appeared before the Board the Bond requirement is waived. Complicated? Yes. If you have questions contact the CSLB or an attorney.

The license exam can also be waived under certain conditions. One way is where the qualifying individual has been listed as a member on the company's Board for 5 of the preceding 7 years and has been actively participating in the construction business. Another way is for the qualifier to be grandfathered in if the business is a family-owned business. This option is only available for corporations, and then only for corporations where a family member in the family-owned business is retiring and another family member who has worked in that business wants to take over so that the family business can continue.

Remember though, that you must have a qualifier and in order to change qualifiers you must notify the CSLB and the CSLB must approve the change. Do NOT forget.

Last but not least, you can joint venture with another licensed contractor, but you need to get a joint venture license. If you don't have the joint venture license, then arguably, you are not properly licensed. The process for the Joint Venture license is fairly simple and straightforward as far as the CSLB is concerned, although, like any other business arrangement, the devil is in the details, and the details are often the joint venture agreement itself. It is important though that you get the joint venture license, so don't forget it.

When Are You Not Properly Licensed?

You are not properly licensed when you either don't have a license at all, or you don't have a proper license for the work you are doing. You are also not properly licensed if the RMO or RME that qualified for the license leaves and you don't replace them, but you continue to do work in that category. There is a 90-day time limit to replace the corporate qualifier — RMO or RME — but you must do it.

And don't be misled; just because the president of the corporation has a license this does not mean that the corporation is properly licensed. If the president of the corporation is not the qualifier (RMO or RME) for the corporation, and the qualifier for the corporation leaves and the corporation continues to contract, then the corporation does not have a proper license. (1) The fact that the president has a license may allow him/her to register as the new qualifier, but register they must. The corporation and the individual are separate; the president may have a license, but the corporation is not the president, it is a separate legal entity. It needs its own license, and the way it gets licensed is through the qualifier.

If the corporate qualifier decides to take off for South America to do missionary work in the Andes for a year or two, and he leaves you the keys to the company truck and the office, is the company properly licensed? (2) No! There has to be some connection between the work the qualifier does and the company. Another qualifier needs to be hired if the original qualifier is out of touch for any significant length of time, or else you, the heir apparent, need to take the licensing test. If you don't, the company is not licensed.

> **NOTE:** For partnerships, if the licensed partner leaves, the partnership can request a 1 year continuance of the license to finish up the existing projects. No new projects can be undertaken until a new partner with a license is engaged, as a partner, in the partnership.

What happens if you don't have the license, but you very diligently hire all licensed contractors to do the work? Let's say you are going to build an office condo project. You buy the land, you find some tenants, and you agree to build the building for them. You then get all licensed contractors to do the work. Can you contract with the tenants to build the project for them even if everyone you hire to do the work is licensed? You can see where this is going. No, you can't do this either. (3)

OK, so last shot, you are properly licensed and your license is current. You fill out your contractor's license renewal and you write that you are exempt under Workers' Comp because it's just you; you don't have any employees. You are representing that you are a paper contractor—you get the jobs and sub them out. Except that you don't; you actually do have employees that you hire for the jobs you get, but you don't want that big Workers' Comp premium forcing you to raise your price or take away some of your profit. Clever idea to a point, right? Wrong!

You are not properly licensed because you have not complied by providing the proper information. You have not paid Workers' Comp when you were supposed to, so therefore your license was suspended all along and you didn't even know it because the CSLB didn't know you were being dishonest, so they could not correct your dishonesty, and voila', your license was suspended as a matter of law the whole time! (4) If you have been providing billings to your clients, and those billings show labor costs, then you have provided the evidence that will prove to be your undoing; and it is a dramatic and painful undoing.

> NOTE: As of January 2011, if you are caught without Workers' Comp., your project can be issued a Stop Work order immediately. There may also be criminal consequences and fines if you do not have proper Worker Comp. on the job. Don't be foolish. Get the proper insurance and make sure you keep it current.

Substantial Compliance Exception

There is one exception to the licensing requirement where you can get a break if you substantially complied with the CSLB rules and regulations. This is not something to rely on, though, because the facts have to be just right for this exception to come into play. For one thing, you need to have been properly licensed in the first place, and the reason you are not currently licensed when the Board catches up with you has to be an inadvertent one. In other words, you were not licensed but you did not know it because of some fact or circumstance that prevented you from realizing that you were not current with your license. Basically you were operating under a good faith belief that you were in compliance, but you weren't, and when you found out you weren't, you immediately set about correcting the problem. (5)

For example, assume you have a license and you keep it in good standing. You move your office and you send the license renewal to the CSLB but you include the old address. Something happens after you move and the CSLB can't get in touch with you and you are blissfully unaware that they are trying. When there is no response from you, the CSLB suspends your license until you correct the oversight (i.e. a contractors bond is not current or you forgot to do the Workers' Comp declaration). You are therefore operating without a license, but you are in substantial compliance and have a good reason to explain why you were not actually compliant.

At one time, Substantial Compliance was a lot more flexible argument than it is now. Don't rely on it to excuse or to try and explain non-compliance. It's there as a way for otherwise compliant contractors to avoid the draconian results that follows for non-licensure, not as an excuse for scofflaws.

What Happens If You Don't Have a License?

If you don't have a license for the entire time that you are doing the work, then you are not compliant. First, you don't have to be paid. You can have done the entire project, provided all of the labor and materials, and done an award-winning job. You still don't need to be paid. You can't sue anyone to get paid either. The courts will throw you out faster than you threw out last week's lunch that you just found in the bottom desk drawer. You have no legal rights. None! (6)

If you don't have a license and have been paid, you can be forced to give back (disgorge) all the money that you have been paid. All of it! It doesn't matter if you have done the award-winning project and Architectural Record is there for the photo shoot; you will need to give back every penny you received. If you are the General Contractor, it will not matter if you paid the subs. If the subs are licensed, they get to keep their money. You, on the other hand, have to pay back all the money, even the money you paid to the subs.

If the Owner has already paid you $100,000, and owes you another $25,000 for some additional work that you did but have not been paid for, you don't get to offset that amount against the $100,000. You still owe the $100K and are out the other $25,000.

What happens if you start work without the proper license, but get the license at some time during the project? You are working without the license for only part of the project, and for the other part you have all of the proper licenses. You should get paid for the work done when you are licensed, even if you have to abandon any claim for the money that you are owed for the unlicensed work you did.

It doesn't work that way in California. (7) If you are not licensed for the entire time you do the work, you don't get paid for any of it. And when they say all of the time you do the work, they mean all of the time. Don't start the work if you don't have the proper license.

> **NOTE:** As an example of just how strict this rule is, in one case a contractor bid the work, got the project, but had not yet started any work. He did not have the license at the time of bid, but that's OK, provided he had the license when he actually started the work. This fellow prepared some submittals and sent in certificates of insurance, but he had not started any work on site. Before he started the work, he obtained the proper license. He then did the job and at the end the Owner wouldn't pay. One of the claims was that he wasn't licensed the whole time he was doing work. Remarkably, the court agreed. (8) The court found that preparing the shop drawings was part of the work, because he could not do the work without the shop drawings. Therefore, he had to be properly licensed when he prepared and submitted the shop drawings, and because he was not, he was unlicensed for the whole time and all the money he was paid had to be given back to the Owner.

The state and the courts are serious about licensing. You should be too. If you are unlicensed, you have no right to make any claims for payment. This does not mean you can't sue for things like discrimination or harassment, because the contractor's license laws do not cover those types of claims. (9) Only contracting and the right to be paid for the work that you do are covered by the licensing rules.

The state was lax for a long time in its enforcement efforts, but the legislature and the courts have now made it very clear that unlicensed activity in construction is not going to be tolerated. If you do not have a proper license at all times when you are performing the work, you cannot come into court to obtain payment for your work. Worse, if you do go into court and you are not licensed, the court will not only throw you out, the court will also direct that you pay back all the money you've been paid for the work. If you have an attorney fees provision in your contract, you can also expect to pay the fees for the other side, too, because you can never be the prevailing party. You will always lose this case.

The icing on the cake, though, is the fact that if you are unlicensed and the Owner knows you are unlicensed, and being the good guy that he is, he agrees that it is OK for you to be unlicensed because he really wants you to do the work, it will not matter. Even if the Owner is in some way complicit in the unlicensed act, you, the unlicensed contractor, will be solely responsible. If at the end of the project the Owner decides he doesn't want to pay you, or worse, wants all his money back, you will have to pay it back.

Therefore, do not start work unless you are properly licensed. There are still a lot of open issues in this particular part of the law, but the one thing that is abundantly clear is that you will not be paid. Worse, you will have to disgorge all money paid to you if you are hauled into court and don't have a license. There are some minimal exceptions to the licensing law, but for professional contractors who make their living in the construction field, these are not especially useful.

> **NOTE:** One issue that is not decided yet is where a joint venture group does not have the proper joint venture license at all times. Are they compliant? Maybe yes, maybe no. If the reason for the license law is to protect against unqualified persons doing construction work, then that reason is satisfied. Both joint venturers have licenses so they are qualified. If the reason is that all contractors must be licensed at all times of the work, and if the joint venture is a separate contracting entity, then without the JV license they are not licensed and the rule is that you must be licensed at all times of the work. Don't be the test case; get the joint venture license. It's easy and it's cheap.

Final Considerations and Exceptions

As we discussed earlier, if your work is less than $500, you don't need a license. If you qualify as an Owner-Builder exception you don't need to have a license either. (10) In this case though, it is not likely you are a true "contractor" anyway. The Owner-Builder exception only applies in residential construction on your own property and for a limited numbers of structures. It has some fairly unique requirements that need to be met for it to be allowed. There will also be an Owner-Builder declaration that needs to be filled out when you go for your building permit.

If you are building spec homes or commercial buildings, you are not an Owner Builder; you are likely a developer. Don't enter into any agreements with buyers where you agree to construct the project for them unless you don't care if you get paid. Remember, you need a license to contract to build things, and it doesn't matter if you hire all licensed contractors to do the work, you still need to be a licensed contractor yourself.

One last point; you may not borrow a license. (11) If you don't have the license, you can't do the work. If you are licensed in another trade, you may be able to subcontract it, but you cannot just borrow someone else's license and do the work. You can bid the work before you have the license, but you cannot perform the work if you don't have the license. Likewise there are certain instances where you need a special certificate to perform the work. Again, you can bid the work, but you will need to subcontract to someone else to actually do the work. Asbestos abatement is one relatively common situation where this comes up.

If you do decide to lend your license to someone, then you are aiding and abetting an unlicensed activity and you can have your license suspended or revoked. There are also criminal penalties that can attach and some hefty fines as well. Unless you owe the guy you are lending your license to a whole lot more than your livelihood is worth, it's not a good idea to lend out your license.

If you end up going to court and come out on the losing end with a judgment against you, unless you pay the judgment or post a judgment bond, your license is suspended. If you or the company doesn't pay the judgment and you are the qualifier on the license, until the judgment is paid you cannot contract or qualify for any other licenses. It means that you do not have a license. (12)

Therefore, if any other company that you are the qualifier for enters a contract while there is an unpaid judgment pending against the first company you qualified for, and there is no judgment bond posted, the other company(ies) is acting in an unlicensed capacity, and we know what that means. If there is an unpaid judgment and no bond posted, any company that you are the qualifier for during this time is acting without a license, so good-bye all payments.

> **Note:** The judgment that is against your license needs to be related to your construction activity. If you don't pay a personal injury judgment, for example, that should not matter unless the PI claim is somehow from or related to your construction work. The judgment must be related to the work you do as a contractor.

Summary

So, are we clear? If you are contracting, you need a license! Your license allows you to do the particular work you are licensed to do. You can do things that are incidental to your classification and specialty, but you cannot do another trade's work unless you have that license classification.

Do not work outside of your classification. Do not work if you do not have a license. Pay all judgments or post a judgment bond. Make sure to stay current with your license; check the CSLB website now and then to make sure they show that your license is active. It only takes a couple of minutes and is well worth the effort. If you are a corporation, make sure you have the proper RMO or RME, and check their license status on occasion. If you are going to work with another contractor on a job, get a joint venture license.

The CSLB and the court's are serious about licensing. You should be too.

Chapter 2

Contracts–Generally Speaking

THIS BOOK PROMISED NOT to make you an attorney, and it won't, but it seems that at least a basic understanding of contracts is important and would be helpful. After all, that is what contractors do; they contract. The problem is that sometimes you think you have a contract and you don't, while other times you don't think you have one, and you do. Tricky things, these contracts.

What Makes a Contract?

A contract, at its most basic, is an agreement. It can be written, it can be oral, and it can be implied. The form of the agreement doesn't really matter; the legal consequence is pretty much the same. Written is usually the most clear (at least it should be), oral is the most difficult to prove (generally), and implied can be the most surprising (really surprising sometimes). So how do these things we call contracts get born into the legal world?

First, there is an offer, then there is an acceptance, and then there is the necessary consideration. Each of these fairly straightforward words is a legal minefield that has been the source of many a lawyers very lucrative practice, because each is an argument in and of itself. And we all know how lawyers love to argue.

What is an offer? Just what it says. It's a proposition; I'll do something for you, if you'll do something for me. Simple idea; very complicated in fact.

What's an acceptance? In the legalese of law school, it's a manifestation of an intention to be bound and perform. Sounds enticing, depending on whose doing the binding and what the performance is all about, but we digress. It's still a pretty simple idea.

And consideration, what's that all about? Well, it's money, for one thing. Or, it's not doing an act that you have the legal right to do or, conversely, doing something that you have no legal obligation to do. Not especially difficult to understand, but it needs to be there or you don't have a contract. In short, it's the reason most deals are done.

Let's take a closer look at some possible contracts and see why they work or don't work as binding agreements, which is, after all, what we intend any contract to be: binding.

EXAMPLE 1. You and I are standing in front of my house, and I say, "I'd love to get this old place painted. Why, I'd pay good money if someone were to paint the place." You don't say anything, but while I'm at work the next day, you show up and paint the place. Do I owe you anything? Probably not.

Why?

Because there was no offer, for one thing. All I said was I'd pay good money to have the place painted. It was not an offer; it was a statement. Was there an acceptance? It doesn't matter because there was no offer, so we don't need to move on in the discussion.

EXAMPLE 2. Let's say I ask if you would paint my house if I paid you $1,000. You laugh, we go in and have a couple of beers, we don't talk about it again, but the next day you come over and paint the house. Was there an offer? Yes, probably so. Was there an acceptance? No, not really. You laughed, and we forgot about it. You painted the house though, what about that? Well, thanks, want another beer?

EXAMPLE 3. I say I'll pay you $1,000 to paint my house. You say you'll do it for $1,500. This time, I laugh and we go in and have the beer. The next day you paint the house. Offer? Yes. Counter offer? Yes. Do I owe you for the paint job? Probably not, but if I owed you anything it would only be $1,000.

The point of this foolishness is to demonstrate that offers and acceptances do not always result in agreements. It depends on how the other party construes the offer, and whether there is really anything to accept. If there is a valid acceptance, what was it exactly that was accepted?

Offer and Acceptance: The Old Fashioned Way to Contract

So what's an offer? An offer is defined as "the manifestation of a willingness to enter into a bargain, so made as to justify another's understanding that their assent to that bargain is invited and will conclude it." Seems like an offer is in the ear of the beholder, doesn't it? Well, yes and no. There is the requirement of certainty as well as a requirement of reasonableness. Not every statement is an offer, even if the other person acts on it. If I tell you I'll pay you a million dollars to paint my two-car garage, I may owe you a million bucks if my garage is the mother of all garages and you paint it. More likely, you just painted my garage, so thank you very much! I may owe you something, but not a million dollars.

However, and we will get to this later, under certain circumstances your offer and my reaction to that offer may very well mean that we have a deal, even if you didn't mean it and even if you can't really perform. This is especially important for subcontractors to know; do not be careless in what you offer to do!

Once the offer is made, the next step is acceptance. How do you accept an offer? First thing you DON'T do is change the terms of the offer. If you accept an offer, you accept the terms of the offer. If you change the terms of the offer, you have made a counter offer and the other side now has to accept your terms. You have switched positions in the negotiations.

The offer also has to be accepted in any way that is required by the offer, assuming, of course, that the manner of acceptance is one of the terms of the offer. Assume the offer is that you can have a really lucrative project, but the only way you can accept the offer is by dancing a hula around a flagpole while someone plays *"It's Only Rock and Roll"* on a bagpipe. If you want the project, then you need to accept by doing the hula, so have at it. It is important that you accept the offer in the required manner, no matter how pointless the ritual is, and no matter how irrelevant the acceptance procedure may seem.

If there is no specified form of acceptance, then typically acceptance is made in the same form as the offer. If the offer comes by mail, the acceptance can be by mail. However, any reasonable method of conveying the acceptance is allowed, provided it clearly conveys that the offer is accepted. Even if you bellyache about what a raw deal you're getting and how you think it's not fair, but you'll do it anyway; that too, is an acceptance. It's a 'grumbling' acceptance, but it's an acceptance. As long as the grumble does not change the terms of the offer or reject the offer outright, a court is likely to find an acceptance. Of course, if there is a gun, literally or figuratively, at your head, then there are other issues that come into play, duress being the most obvious, but more on this later.

The California Way: Mutual Assent

In California, the courts don't rely too much on offer-acceptance type of analysis but rather on "mutual assent." It's basically the same thing as offer-acceptance, but the courts analyze the circumstances of the agreement with less differentiation between who did the offer and who did the acceptance. Instead, it's more of a situational analysis; did the parties know what they were doing, and intend what they were doing, when they did it?

The court looks at all of the facts to confirm that the parties understood and intended what the contract says they will do. It's very much an objective analysis. It won't matter that you didn't really understand the finer points. You can't pretend you didn't know what you were doing if objectively it would seem that you did, or that you should have understood. The court won't undo a contract except for very good reasons. Just the opposite; courts will try to find a contract and give effect to the mutual intentions of the parties.

Of course, if there is a mutual mistake of fact or law, you can probably unwind the deal, but that's a tough proposition. If you make a mistake of fact and the other guy knows you are making a wrong assumption, but he doesn't say anything, you may get out of the deal. The terms need to be clear, and if you know something that would materially affect the other side's willingness to enter the deal, you need to disclose it if the other side would not know it otherwise. If the other side is just lazy, and should know it but didn't look into it, well, tough! A deal's a deal.

Capacity

This raises the issue of capacity to contract. If the person is mentally incapable of knowing what they are doing, then the deal is going to be undone. Likewise, if the person is not legally of age to contract, the deal is not going to stand. You are not going to have an agreement with a middle school teenager to sell you his classic fully restored '66 Mustang 2+2 for $500 stand up in court. Finally, if the accountant for the millionaire signs your company up to build the client's multi-million dollar home, you should probably check with the millionaire client to make sure the accountant can actually enter into such an agreement on behalf of the client.

If there's a Remedy, there's Likely a Contract

One way the courts decide if there is a contract is to determine what it would take to breach the agreement and, if there were a breach, what would be a remedy for that breach. In other words, if one party fails to perform, what would be a remedy for that failure? If there is a remedy, then there is likely a contract. In the construction world, contracts are not hard to find, because there is typically a remedy for any breach. Whether that remedy is fair or not is another issue, but there is usually no problem determining if a contract exists.

Consideration

The last element for contract formation is consideration; some form of benefit or detriment that both parties obtain by the agreement. It can be money or it can be services, or it can be a combination of the two. Consideration can also be an agreement to do something, or to refrain from doing something, in exchange for the other party's promise to do, or not do, something. For example, I will paint your house if you don't take me to small claims court for the $200 I owe you. I don't have to paint your house, but I will, and you can take me to small claims court, but you won't. We both contribute something to the bargain; if I paint, then you won't sue.

Things That You Can't Contract to Do

Contracts don't always get formed, even if they meet, or appear to meet, all the above conditions. First of all, the contract can't be for something illegal; i.e. I'll pay you $10,000 if you bump off Joe. Or, more likely, I will pay you $5,000 if you get me that City job. Bribery, extortion, and the like are not proper bases for contracts.

A contract can't be for something that you already agreed to do; that is, a pre-existing duty. For example, if you and I have a contract that says you will paint my garage for $200 but then you demand another $50 for the job for no reason other than you want another $50, I can promise to pay it, but I don't have to pay it, because you already have a pre-existing duty to paint the garage. If there is nothing more that you have added to the agreement (no additional benefit), a court is unlikely to enforce the second agreement for the added $50.

What can be enforced, though, is if the $50 is consideration for another agreement; i.e., you'll paint my garage in the next week rather than when you get some time between your other jobs. Or if there is some added work or additional costs that are incurred for supplies (I change the plain white color to a custom color), then we have a new and valid contract. There has to be some new consideration from you for the added $50 payment from me.

Another non-contract arrangement is where Aunt Molly says she's going to give you $500 next Thursday. She doesn't, though, and so you want to haul her into court and get the $500. Not likely! It was a gift and the gift can be revoked at any time. Therefore, if your favorite sub promises to give you that chain saw you really like, but then decides not to give it, there is no way to compel him to do so as long as it remains a gift. If you did something in return for his promise of the chain saw, that's a different story, and you do have a right to compel him to give you the saw.

So there are contracts and non-contracts, depending on the circumstances. The contact to kill someone can't be enforced, at least not by a real court; the pre-existing duty rule gets me out of the extra $50 to paint my garage; and the gift is a gift and not a contract by its very nature.

Promissory Estoppel/Detrimental Reliance

Recall that earlier we said you could be in a contract and not even know it. This is one of the ways that can happen. These are important concepts, and you need focus here. The concepts are not difficult, but if you don't understand them and get caught up in the intrigue they create, it can really wreak havoc.

Promissory estoppel is a 'promise' that you make that you are 'stopped' to deny if the other side relies on it to their detriment. Not the most eloquent way to put it, but hopefully easy to remember by the sound shifting that was used; 'promissory' for promise; 'estoppel' for stopped. So what does it mean in the real world of contracting?

A lot, especially in the world of public contracting.

If I say I'll paint your house for $200 and nothing else is said, then as we saw earlier, there is no contract. It was an offer but you didn't accept, and so I don't have to paint the house.

Change the situation a little. Let's say you are looking for someone to paint your house and you ask for bids and I am the low bid at $200. I change my mind and tell you I can't do it because I need to go to Vermont to see the Fall colors. Contract? No, probably not; at least not one really worth pursuing. In all likelihood you will negotiate with one of the other painters you contacted and get them to do it.

Change the situation again. Let's say there is a bid for a project that requires the painting of a house and you get my bid for $200 and use it to make your bid for the project. You end up as low bidder and get the project. I then tell you I am off to Vermont and can't do the job.

Do we have a contract? Yes, maybe. It depends on whether you relied on my bid and whether your reliance was reasonable. It's fairly obvious you relied on my bid, but was it reasonable?

If my bid of $200 was in the ballpark of all the other bids to paint the house that you received, then, yes, it was likely reasonable to rely on the number. But if my number is off by a significant amount and it is obvious that I made a mistake, then no, you can't hold me to the number. If all the other painting prices were $20,000, so it was clear that I forgot a couple of zeros, then it is not reasonable at all to use my number, or to rely on it. Plainly something is wrong with my number.

If the number I give you is close to the number you get from others, then my number is likely reasonable and you may be able to prove a claim against me. If the number I give you is lower by some fairly large amount, you may still be able to say it's reasonable to rely on the bid. For example, if my specialty is painting houses and I do lots and lots of houses and I also own a paint manufacturing company, and you and I have done business in the past and my price has been consistently lower than my competitors on those other jobs, and we do a lot of jobs together, then yes, you have some good reasons to rely on my bid. Maybe not a bid that seems to be missing a couple of zeros, but at least a number with the same amount of digits.

PE/DR in the World of Public Contract Bidding

In public works, the bid form submitted by the General must include the name of any subcontractor whose bid is ½ of 1% of the total bid amount. The bids all go in at a specified time, and the General Contractor has to post a bid bond that gets forfeited if the General will not perform once the public agency determines the General is the low bid. There is only a very limited time to pull a bid, and it can only be pulled for very specific reasons. A sub's mistaken number is not such a reason.

If the General Contractor gets your bid and lists you as the subcontractor for that scope of work, you are on your way to a lawsuit if you try to pull your bid after the General submits his bid to the Owner. There are no 'amendments' to a public works bid after the bid is submitted. (All of this will be discussed in more detail later, but this thumbnail is to give you a sense of the seriousness of the bidding process, especially in public works.)

The point here is to show how promissory estoppel/detrimental reliance functions in the real world, especially in the real world of public works contracting. The General has listed you; she has relied to her detriment on your number and you are bound to perform for that number. Unlike the typical give and take of the offer-acceptance-counteroffer mutual consent arrangement we discussed earlier, in this instance, the law implies an offer by you that is accepted by the General when the General uses your number in her bid to the public Owner. Assuming it was reasonable to rely on the bid in the first place, if you try and pull your bid, the General will be harmed, and it will be construed as a breach, notwithstanding that you do not have a written agreement at the time.

Efficient Breach

What happens if you really can't do the work for the price you gave? The General will need to find a replacement contractor, but that replacement is likely to be more expensive than you. As a consequence of the breach, you will pay the difference between your bid and the replacement contractor's bid; this is the General's damages.

Does that mean the General can get her boyfriend's company to come in at a price 3 times your price and you have to pay the difference? No. The replacement cost has to be reasonable, and it will be. You'll do the job if you will lose more money by not performing than you will by doing it at the mistaken price. This is the legal concept of efficient breach; when it is better for you not to perform the contract than to perform it at a loss. You may pay damages, but the damages will not be as much as you would lose if you did the project. The General also has the obligation to mitigate its damages by acting promptly to get a replacement and to get a good price from the replacement contractor.

How Long is the Bid Good For?

How long do you need to hold your bid price? The short and non-answer is, a reasonable time. What does that mean?

Let's use the public works project again. The bid documents provide that the Owner intends to award the contract within 60 days of the bid opening. The bid documents require that bids must remain open for that 60-day period. The General has to keep her bid open for that time frame, but does that mean the subs have to hold their bids open for 60 days? Can the sub condition its bid on being awarded a contract in 30 days? This is a tough question, and it can generate a lot of attorney work.

The way this is typically argued is that if the sub's bid proposal for work on the project is per the plans and specs, then the sub should be bound for the same time frame that the General is bound. Otherwise the General really has no assurance that it can do the work for the price that it quotes on bid day. All the subs could limit their proposal's acceptance period, and when that time frame passes they could increase the price or just refuse to do the work. The initial proposal would be worthless. Therefore, logically and legally, the better reasoning is that the General's bid and the subcontractor's proposal run concurrently and are open for the same time period as is provided in the Prime Contract with the Owner.

But let's say the sub provides a bid, the General uses it, the General gets awarded the project, but then the General does not issue the subcontract for the work until months afterward. In the meantime, the cost of materials has gone up, the sub has not been able to lock in its costs because it has no contract, and when the General finally gives the sub the contract, the cost for the sub to do the work has increased significantly. Can the sub refuse to sign the contract? Can the sub change the cost of the proposal that was provided? It would seem the answers should be yes.

The General should have sent a letter to confirm that the subcontractor has the job, that a subcontract would be issued, and that the sub should lock in its prices if it has not already done so. If the General doesn't send such a letter, then the subcontractor has a strong argument that its bid is no longer effective. The General had a right to rely on the validity of the bid for a reasonable period, but the General cannot interfere with the sub's ability to perform by unreasonably delaying the sub's ability to secure its costs and its performance, and, after the unreasonable delay, still expect performance.

This last hypothetical is another example of promissory estoppel; the sub is relying to its detriment on the General's promise to use the sub on the job (the sub was listed after all) and therefore, if the costs go up during the unreasonable time that the General takes to get the subcontract out, or to at least send notice that the sub will get the subcontract for the work, then the sub has damages that the General arguably may be responsible for.

Alternatively, the sub has a good reason not to perform the contract because it is not reasonable for the General to waste so much time getting the subcontract out. One of the implied warranties in all California contracts is the implied warranty of cooperation; we can't stand in each other's way as we do the work, and that includes issuing the subcontract.

PE/DR in Private Works

Given the less restrictive environment that exists for private works, the really difficult questions of promissory estoppel and detrimental reliance don't usually come up. That's not to say they don't; they do. One way the issue can come up is where you have the opportunity to do a really big project, but you only have limited crews. The project is going to take all your time and all your attention. You've provided your price and you've told the Owner that if you get this job you are not going to take any other work until the project is complete.

The Owner likes you and your crew, and tells you that you have the job and that he will get you the contract just as soon as he gets back from France and his buying trip for the project. In the meantime, you finish up all your other work and don't bid anything else because you're busy getting ready for the big project.

Then the Owner comes back from his trip and he tells you that he has had a change of heart and wants to use someone else. You have no contract, but you do have your bid and the communications that went between you and the Owner telling him how you are really looking forward to this project and how you will not be doing any other work because you are devoted to this job. Do you have a claim?

Yes.

You relied to your detriment on the Owner's assurance that you had the job, and it was reasonable to do so. You are damaged because you turned down all the other work that was available and, while you are diligently looking to get some other work, you are now behind the proverbial eight ball. Your damages are, at a minimum, the profit you would have made on the project if you had been able to do it. Your proposal will show your estimated profit and the Owner will have understood that you anticipated that profit. You should get that amount for the breach, and possibly more. After the attorneys get through with it, you may not have much of a profit left, but in the private context, that is one way promissory estoppel can be applied.

Summary of PE/DR

With promissory estoppel/detrimental reliance, the point is to be aware of what you bid or, conversely, what bids you accept. If you accept the absurdly low bid and use it to get the job, a lot of the profit you included will go to paying the difference for the replacement contractor you will need to hire and pay. If it's unreasonable to take the bid, you will not get damages. Conversely, if you give a bid that is reasonable, but you don't want to do the project, you will pay the difference between your price and the price of the replacement contractor.

Promissory Estoppel/Detrimental Reliance is all about fairness. You should not suffer damages if you act reasonably and in good faith and the other side does not. It is a way that contracts are formed without anything in writing. Just the fact that you offer to do something, or the other side relies to its detriment on your offer to do something, and assuming it was reasonable to do so, will create the contract. The damages for the breach of this type of contract situation are the same as for any other contract; whatever it takes to make you whole and give you the benefit of your bargain.

Breach: What It Is and What You Can Do About It

Once you are in contract with someone, and you, or they, decide that you don't like the deal any more, or one or both of you decides you don't want to do the work, what happens? First of all, it depends on what kind of contract you have. If the contract is for goods (i.e. materials, oranges, tutus, that sort of thing) then you have different rules to deal with as opposed to contracts for services (framing, plumbing, bulldozing new roads in virgin old growth forests). If you provide goods, you need to look at the Commercial Code for what constitutes breach and what you can do about it. If you are building something, or if you are anyone other than someone that sells goods, you follow the Civil Code and Common Law, which is what we are talking about in this book.

So what are your options if the other party decides not to do what they promised to do? If the breach is a material breach, one that goes to the very essence of the agreement, you have three options.

- You can stop your performance and immediately sue for your damages.
- You can continue your performance and then sue for your damages.
- You can treat the contract as rescinded, so that you no longer have any agreement at all.

Two important things to keep in mind with breach of contract:

1. You must give notice to the other side that they are in breach, unless it is so obvious that the other party is not performing or is not going to perform; and
2. You must have damages. There can be no breach without damages however that term may be defined.

How do you know whether you can stop performance or, more important, terminate the contract? It's tricky, and it's risky, so you do not terminate a contract impetuously. You need to be very, *very*, careful, because if you are wrong, then you are the one in breach and the other side gets its damages — from you!

When Can You Terminate a Contract?

First, there is the legal concept of substantial performance; that is, performance that has effectively achieved the goals of the agreement, except for some incidental and minor issues that may not be completed. A common example is the punch list. If you are done with the work except for the punch list, or some final 'to-do' list of the Owner, then you have substantially performed. There can be no material breach if the contract is substantially performed.

This does not mean that if the only issue on the punch list is the HVAC unit that cools the entire building in Palm Springs, but the HVAC is still not installed and won't be for some time, that the contract has been substantially performed. Likewise, if the punch list consists of thousands of items that need your attention, then you are probably not done and you likely have not substantially performed.

The point is that the scope of the punch list and its impact on the actual use of the project will determine whether there is substantial performance. If there is substantial performance, then the damages for any breach are going to be limited to the cost to perform whatever work remains that is not complete. There can be damages, but the contract can't be terminated, except under some really extraordinary circumstances.

Second, the type of breach will determine if you have the right to terminate the agreement. If the breach is a material breach, something that goes to the essence and purpose of the agreement itself, you can stop further performance. In order to determine whether it is a material breach, though, the terms of the agreement will become critical. What, exactly, was agreed to when the contract was signed? Is the issue that is now becoming a problem something that was expressly addressed in the contract, or is it so obvious as to be simply assumed, or is the particular issue left un-addressed?

For example, if the agreement required the Owner to provide certain critical materials by a specific date and those materials are nowhere in sight, this is potentially a material breach. Potentially, because if you can work around the missing materials, then you must. But if you get to a point where the critical materials cannot be ignored anymore and the job comes to a halt, then this is a material breach and you are entitled to damages. You also have the option to terminate the contract or suspend performance.

The important point to understand is that just because the contract says that one party will do something, but they don't, you can't just abandon the project. You must continue with your performance until the other party's action undermines the very basis of the agreement and you are not getting what it was agreed that you would get. Be very careful with terminations of contracts, because if you are wrong about the breach and you terminate or cancel the contract, you will be the one in material breach.

Determining what is material to the contract, if it is not spelled out in the agreement itself, is not an easy determination. You will have your view of the matter, the other party will have theirs, and the court or jury will have its own. If the option you intend to exercise for the perceived material breach is termination of the contract, then take a deep breath and think it over very carefully, not just from your perspective, but from the other side's as well.

Termination should be the very last option you consider, and it should only be done after you and your attorney have done everything you can to get the other side to perform. The events leading up to the termination should also be well documented in your project files to show that you have done whatever you could to avoid the ultimate decision.

Other Contract Situations

There are some less typical situations that sometimes come up in construction projects. Let's say that you have a contract that says you will have the project done in 9 months, and you will be paid for the entire project once it is complete. Assume it is an elegant stairway, made of exotic woods that you will construct in your shop from materials that you will hand-select. When it is done, you will disassemble it and reinstall it in the client's home. It's not an unlikely situation, and it's helpful to illustrate the following points.

Being the temperamental artiste' that you are, you find yourself distracted by other things during the course of the 9 months that you are supposed to be working on the stairway. You are just not in the mood, you don't feel the spirit, the vibes are not right, you are not one with the material and it is not speaking to you; all that sort of thing.

The client calls you around the 7th month and asks how it's going, and you tell him. He is, needless to say, not impressed, because this stairway needs to be done so that the rest of the palace can get constructed around it. You've frittered away 7 months and not done anything.

Situation 1: ***Time for Performance.*** The Owner tells you that you are in breach and that he is canceling the contract. After you slam the phone down repeatedly while telling him to do what is physically impossible for a male to do to himself, you realize, "Hey! He can't cancel the contract!" And you would be correct.

Why? Because you are not in breach yet. The time for performance is nine months and it is now only seven months. You are not in breach until the time for performance has passed and you have not performed, or substantially performed. You call the guy up and tell him if he terminates the contract you will see him in court and you will sue him for breach of contract. We will leave it to your attorney to figure out your damages, but the point is that if the time for performance has not yet arrived, there is no breach.

Situation 2: ***Specific Performance.*** You call the Owner and tell him that you don't want to do the project. You're just not into it, and you will give the guy back whatever retainer he may have paid, and if he wants the exotic wood you found on your trek in Tibet, he can have that, too, if he will pay for it but, you're just not interested anymore.

Can he make you do the job?

Yes, in all likelihood, he can. He can sue you for specific performance. This is a relatively limited remedy that requires that you have some special or specific talent or ability that no one else has, so that you, and only you, can do the job. Only Rembrandt can paint a Rembrandt, so if you and he have an agreement that you, Rembrandt, will paint the picture, then he can likely get you to do it, if you have a contract. The staircase is unique, and because it is one of a kind and only you can build it, you will likely be required to build it.

SITUATION 3: ***Adequate Assurances.*** The Owner calls seven months after you sign the agreement to find out how you are doing and, as before, you still have not started. This time though he is a bit more savvy. He knows he can't terminate the contract because you still have another two months to get your act together and build his stairway. What can he do?

He can send you a letter demanding "adequate assurances" that you will be done in two months as agreed. He is entitled to an answer, too. This is called anticipatory breach in the legal parlance and often occurs in contracts for the sale of goods, not services, but it is something to consider if you have doubts about the other party's willingness or ability to perform.

If you do not respond in the time frame requested, then, in this example, the Owner may be able to terminate the contract and then sue for damages based on your anticipated breach. If you write back and tell him that the stairway will be complete as agreed, he can't really do anything. Of course, if it is obvious that you cannot perform in the two-month time frame because it is physically impossible to do so, then he can call it a breach and you can be liable for damages.

SITUATION 4: ***Cardinal Change.*** Assume now that you have started the stairway in a timely manner and you are proceeding and making good progress. The Owner's wife has too much time on her hands and too many designer friends at the country club. She calls and asks if you can make a small change to the plan; change the width of the tread by 2". You agree. Then she calls and asks if you can make the stair rail a different wood. You agree. Then she asks if you can put metal pickets instead of wood pickets in the rail. You agree. Then she asks if the stair can be on the right side of the entry hall instead of the left. Then she wants it to be a circular stair rather than against the wall. Then she says change it to a self-supporting stair in the middle of the entry. Then she tells you to make it out of old growth redwood, not the Tibetan hardwood you have. Then she says... and then you say, Enough!

Can you terminate the contract? Yes.

Why?

Abandonment by change. Actually, in this situation, the correct term would be cardinal change; a change of such significance that the original contract is no longer even recognizable because the changes have completely obliterated the original agreed project. You no longer have an agreement, because the object of the original agreement is not the object of the changed agreement.

If you have been making these various changes as they were requested and the Owner insists that you need to keep your price, or if there is a dispute about the price because of all of the changes, then what should happen is that you get paid for the value of the work, which is presumably a lot more than what you originally estimated. The legal term is quantum meruit, which is a fancy way of saying you get paid for all that you did and the time that it took you to do it.

> **NOTE:** The theory of cardinal change and quantum meruit will not apply in public works contracts against a Public Owner. If you are a subcontractor, you can make the claim against the General Contractor, but if you are the General Contractor you cannot make such a claim against a Public Owner. More on this later, but for now be aware that these two theories can't be raised against a public entity.

Statutes of Limitation

These are critical time limits that apply to every issue that will likely be litigated in construction cases. As we go through the various claims and causes of action that can be raised, there will always be a statute of limitations issue for each. There are three statutes of limitations that you need to keep in mind:

- The two-year statute for breach of oral agreements
- The three-year statute for negligence claims
- The four-year statute for breach of written agreements.

A statute of limitation is the timeframe during which you can bring your claim into court. For example, if you had a written contract on a project that was completed in 2002 and you wanted to sue for breach of that agreement, you have to file your action by 2006, or within 4 years of completion. If you don't, and while there may be other ways for you to get into court, as a practical matter you will not be allowed to pursue that particular claim. If you file your breach of contract action in 2008, it would be too late and the court would throw that part of your case out. You need to act promptly and within the limitations period, or you will lose the opportunity to act at all.

Issues Related to Drafting a Contract

A contract is supposed to be the result of a negotiation where both parties have an opportunity to make offers and counteroffers, so that eventually both sides are happy with the deal. The contract itself is a written summary of the agreements reached during that negotiation. It doesn't always happen that way, and sometimes a "negotiation" is in name only.

Contracts of Adhesion

If one side is at a disadvantage and has little, if any, opportunity to negotiate the terms, and is left with a take it or leave it option, and with no real option to leave it, then what we have is a contract of adhesion. Where one side has all of the negotiating power to decide what is in the agreement and the other side has effectively none, then the contract that is formed may be voidable if the terms are so one sided that they are patently unfair. A voidable contract means one party has the option to walk away because the contract is unfair or otherwise unenforceable.

A void contract means it is not a contract at all. It is an important distinction.

In the construction context, one form of a contract of adhesion is a contract for a public works construction project. (Public Owners may not agree that public works contracts are contracts of adhesion, but, reasonable minds may differ, as they say.) Certainly a General has the right not to bid on the work if they don't like the terms but, if they do bid it and if they are the low bidder, they are more or less stuck with the terms of the Owner's agreement.

This is not to say you cannot change the terms after the contract is issued, but it is unlikely that you, or any other bidder, will get the Public Owner to change the terms of agreement before the bid, unless it violates a statute or is simply void as a matter of public policy. If you can't change the terms before the contract is bid, it is unlikely you will change them afterwards. It is also unlikely the Public Owner will change terms in its contract, because those terms are included for the Owner's benefit, not the contractor's.

Ambiguity in Contracts

If the terms in a contract are ambiguous, that is, capable of two equally plausible and reasonable interpretations, then the ambiguity is construed against the drafter. This can be a very effective defense in disputes, especially with a Public Owner, because it is very often the case that public contract terms are ambiguous. Coupled with the argument that a public works contract is a form of a contract of adhesion, the public agency's contract can sometimes be turned against it in a dispute.

The Long and Short of Negotiating The Contract and Contract Terms

If you have the opportunity to negotiate the terms of a contract, there are some key terms that will always be included and some terms that need to be looked at and negotiated. This is especially true in subcontracts. Depending on the General-subcontractor relationship, the negotiations can be completed fairly quickly if the parties have done business before, or they can be long and hard-fought if the parties have not worked together. Never, never, never, sign a document without looking at all the provisions, and never be shy about crossing out things that are not acceptable or adding things that you want included. When you strike out or add language, always initial the change.

Contracts can be very short or they can be extremely long and detailed. Keep in mind that the more detail you include, so that it seems like you are trying to cover every situation, the more likely a court will be to find that if something is not mentioned in the agreement, it is not covered by the agreement. The reasoning is that you included everything that you wanted, so you must not have wanted what was left out.

Conversely, if the agreement is very cursory and short, i.e. vague, then a court may say that the contract is not inclusive at all and was simply intended to set out the basic terms and general understandings of the parties. This lack of detail may allow you to bring in additional information about what was intended at the time the contract was entered, but so can the other side. The court then gets to decide what your contract means, and that may not be the best thing.

Having a contract that is clear about its purpose and that includes certain key terms is the best. If the critical items are written into the agreement, then the parties can sweat the small stuff later, if they need to. (More on what should or must be included in a contract will be explained in the next chapters.) This is not to say that the agreement should be sloppy or unclear. It should be anything but. If you are drafting a contract, plain words and simple sentences are best. Stay away from legalese; the *wherefore this* and the *heretofore that* type of nonsense. Even attorneys hate this kind of silly writing.

Make the agreement clear, make it concise, and make it complete. If there is a key requirement that is important to you but that the other side may not expect, put it in bold or make it a larger font size so that it is noticeable. If it's not an expected term, it needs to be emphasized so that there can be no question that it was read and agreed to when the agreement was signed. Make sure all parties initial all pages of the agreement. This shows that everyone had the opportunity to read the document; whether they do or not is up to them.

And, of course, remember that if you draft the agreement, any ambiguity is construed against you. Include a provision in the agreement that states that all parties have contributed to the creation of the document and therefore no one party is deemed the drafter of the document. This is one way to avoid the ambiguity issue. Then again, if you don't have any say in the preparation of the document and this type of provision is included, cross it out. If you didn't have any chance to negotiate the terms, don't agree to waive the ambiguity issue.

When you get a contract from the other party, the first thing you have to do is READ IT. Just because you have done business with the other party for the last 10 years does not mean that you can be sure that there will be nothing objectionable in the next agreement, or that you will continue to have the same great working relationship you have experienced in the past. Circumstances change, people change, contracts change. Read the document.

Read the whole document. If the contract is in a small font, read it with your finger; that is, take your index finger and point to each word as you read it. If you read it this way, hopefully you will read it carefully. Small print is difficult to read and it makes a person tired; tired readers are lazy readers; lazy readers miss things. Certainly there are things that you will skip or gloss over, and that is probably OK for some things, but not all things. There are some important sections that you will typically know that you need to read closely, but if the agreement is long and wordy, you need to be especially vigilant and attentive. It's better to catch the objectionable term up front, when you can still discuss it, rather than later when it is an issue in dispute.

To repeat and summarize, when you read the agreement, if there are handwritten cross outs or interlineations, read them carefully. If you make a change, initial it. If you are reading a change and are OK with it, initial it. If not, cross it out and initial it. When you are done reading the final draft of the agreement, and it is acceptable and you will sign it, initial every page. READ IT again; use your index finger to focus your attention. Make sure all pages are numbered. Make sure you date it when you sign it.

If the agreement is still in draft form and there are changes being made by both sides, and terms are still being discussed back and forth, there are a couple of ways to exchange those revisions during negotiations. If you have a good working relationship with the other side and not too many changes are expected, WORD's "*Track Changes*" feature is a great way to see the other side's changes and to keep track of your own. Always Save your changes and always save the changes you receive from the other side. Use the "*Save As*" feature, but make it a different version number. Include the date with the revision number, so that you can track when the various changes were made.

If you don't know the other side very well, or if you want an alternative and less clumsy way to negotiate numerous or long changes, use a separate sheet with just the changes written out and identified by paragraph and then exchange these. Once the changes have all been agreed to, and assuming you have the original document in WORD, you can cut and paste the section(s) you want to change into the final document. This way all that is exchanged are the requested changes, not the entire document. You will be able to read the changes that are made, whereas if you use "*Track Changes*" for a number of iterations, which are all saved, after a while you may not be able to read anything.

Once you've got the finished document ready for signature, and if you are emailing it to the other side for their signature, always make it a pdf and send it that way with your initials at the bottom of each page even if you have not actually signed it yet. You do not want any inadvertent change being made to the document, which is possible if you send it in WORD.

Finally, one way to make sure both sides have an original and complete wet signed copy is to have two copies of the final document made up. Send both to the other side for signature and have them sign and send both back to you. Then you sign and send one of the two Originals back to them. This way you both have, or should have, a copy of the same wet signed Original Document.

Contracts are legal documents with enormous consequences. Do not take them lightly, and be sure you understand what it is you are signing before you sign it. The contract is the way you will be conducting business with the other party until your project is complete. It should be as clear as it can be about that procedure.

Chapter 3

The Contract: Basic Terms and Requirements

In private works, more than public works, there are a number of ways that the parties to a contract can reach the point of entering into the agreement. In public contracting, the Public Owner is typically required to accept the bid of the lowest responsible bidder. This is not the case with private works contracting. In private works there can be an open bid where the Owner accepts prices from a number of contractors, or the Owner can chose to negotiate with only one or two contractors. Or, there can be a combination of the two; put the project out to bid, select from among those bidders, and negotiate with them. The project can be bid or negotiated as a lump sum, a cost plus, a cost plus with a guaranteed maximum, or any other arrangement the parties chose to use.

Under any of these arrangements, the contractor is required by law to provide the Owner with certain information in the contract form. Depending on what kind of project is being constructed, there may be certain language that needs to be, or should be, included. For example, a home improvement contract will have different required language than a commercial contract for a strip-mall shopping center.

The basic contract terms are the same, but failing to include language that is required by statute can be a very bad thing indeed. Therefore, it is important to know what kind of project is contemplated because that will determine what provisions need to be included in the agreement with the Owner.

Form Agreements

There are a number of form agreements that you can purchase at your local office supply store or that are available from various trade organizations; the American Institute of Architects (AIA), Associated General Contractors (AGC), and others. These contract forms for Owner-Contractor or Contractor-Subcontractor under various arrangements (lump sum, cost plus, etc.) are usually very good and can be modified to suit the purposes of the particular project, but only to a point. If the project is very complicated or unique, in the sense that it involves terms and conditions, obligations and duties, that are somewhat out of the ordinary, then it is probably best to draft a new agreement that will address those circumstances. If nothing else, it gets all the parties to think about what they are going to be doing and how they are intending to do it, and in that process any number of misunderstandings or ambiguities will become evident. It is always better to work those issues out ahead of time rather than while the project is under construction.

What should not be done is to take an off-the-shelf agreement and modify it, or worse, sign it because it purports to be a "construction agreement for services." It may very well be that, but it may be an agreement for services in Bulgaria, not California, and so you may have terms that cannot be enforced here or you may have omitted terms that need to be included because of the particular type of project you are going to be constructing.

> Note: This same admonition holds true for any off-the-shelf forms for Mechanic Liens, Stop Notices, or conditional or unconditional lien releases for progress or final payment. Make sure that the language in the agreement is appropriate under California law. Otherwise, you may have compromised your rights in some manner. For contracts, especially, it is a good idea to have an attorney look over any form agreement before you start using it for your projects. It will be money well spent, and it should not be that costly. And make sure the attorney is familiar with construction law and

construction generally. The guy may be an attorney, but he may not know beans about what needs to be in a construction contract.

Pre-Contract Meetings

Before you enter into any construction agreement, you will have looked at the plans and specifications and all the reports, and will probably have visited the site and met with the Owner and design team. The discussions and meetings will have been about various issues and requirements on the plans and in the specifications. If you already have subs involved, then you may have had meetings with them, as well, and they may have had questions for you, the Owner, and/or the design team.

It is imperative that you keep good, very good, records at this period of time. If you have meetings, keep notes and make summaries of those notes and send them as confirming memos to the Owner and the architect. If you have questions, write them down and get written responses. If there are changes to the specifications or the plans, note them in the specifications or on the plan sheet, and obtain confirmation of those changes in writing from the architect. Document, document, document! No matter how great your relationship is with the Owner, and no matter how long you have worked with the architect or engineer, do not trust that your understanding of the discussion is also their understanding of the discussion.

The most trivial and insignificant point may become an incredibly contentious issue when the time to perform the work arises. The fact that you thought it was agreed that the wall was going to be moved 3", and you laid out all of the footings and rough-ins accordingly, will wreak havoc on the project and in your relationship with the Owner when the plans do not get changed, and because you moved the wall 3", the windows and finishes do not work and the plumbing vents don't line up.

Make sure the changes are not just documented, but approved. Written meeting minutes and written responses to requests for clarifications are critical. It's also important to be persistent. If the response begs the question or is vague, ask it again. Ask it until you get a real answer to the actual question. Keep a log of the questions and keep a notebook of the changes. Paste or tape the changes over the existing drawings or next to the detail that is at issue. You want an accurate and reliable record of this time and of these discussions.

Recall, too, that if you get a response to any question, it must come from the person authorized to respond to the question. The assistant to the assistant to the architect likely has no authority to respond to your request, and if they do, because it seems so obvious and straightforward, it will not be binding and you cannot rely on it. Ask for confirmation of the response, or confirmation that the person responding has authority to respond, or both.

When you finally submit your bid, or when you arrive at a price, the documents that you used to get to that price need to be preserved and identified. In your bid, and in the contract as well, all the documents that were used to prepare the bid, including their dates and revision numbers, should be identified and listed. The plans and spec books should be marked as the Bid Set, and they should be set aside and kept for future reference. Before marking them as the bid set, a copy should be made so that you have a job set of the same document. Likewise, the correspondence and meeting minutes of any pre-job discussions should be kept in a binder and preserved. If there are problems down the line about the scope of the work or what was intended, these documents will become invaluable.

The Written Construction Agreement

The basic information is what you would expect for any construction contract in California. There should be a heading that identifies what the document is; i.e. HOME IMPROVEMENT CONSTRUCTION AGREEMENT or CONSTRUCTION AGREEMENT, or some such identifier. A Home Improvement project is, for all intents and purposes, a remodel of some sort, but it can also include adding swimming pools, guesthouses, and/or other secondary structures. Home Improvement is not a commercial building, industrial project, a new custom home, or subdivision type of work. If the custom home or subdivision home involves a "remodel" afterward, that is a home improvement.

In any agreement, the parties to the contract need to be identified. If there is a construction lender for the project, the lender should be identified. The addresses of all should be included, but especially the address of the Owner. The name and location of the project should be included in the opening paragraph. The form does not matter, the substance does; it is the information that is important.

You also want to include a description of what the contract is about. What, exactly, is this contract for? This can be as brief or as detailed as you want it to be. However, given that you will want to include by reference the plans and specifications, the bid proposal, the description does not need to go into especially great detail. As long as someone unfamiliar with the prior discussions between you and the Owner can understand what you and the Owner agreed on, the contract is probably clear on the terms.

If the contract is a subcontract agreement, the terms should include the specification section(s) that the sub will perform and it should note the plans and specifications for that work by date, and include any addenda that may have been issued. If the sub's proposal is to be included, then it should be included by reference. If there are exclusions to the subcontract work, make sure they are specifically noted, or are noted by specific reference to some other document, i.e. the sub's proposal. After the identification of what is included as the subcontractor's work, there should be a disclaimer such as, "No Other Exceptions Taken," so that it is clear that what is described as being excluded is all that is excluded. If it is not excluded, it is included, so beware.

For the Owner-General agreement, something like the following language is sufficient to create a binding contract. It can be modified to work for a Subcontractor-General agreement. What is needed is the acknowledgment that the parties are entering an agreement for performance of a specific project, according to specifically identified documents and plans, and that the agreed price, subject to the remaining provisions of the agreement, will be paid. Again, this is basic information, and it is the formal representation of the offer, acceptance, and consideration that will establish the basic elements of the contract.

> **The following is a Construction Agreement between ________________, (Contractor) and the Project Owner ____________ (Owner). This agreement is a binding contract between the Contractor and the Owner whereby the Contractor agrees to build and the Owner agrees to pay the Contractor to build the following Project:**
>
> **Project Description:______________________________**
>
> **Per Plans & Specifications Included and identified as _______**
>
> **Project Address: ______________________________**
>
> **The Owner agrees to pay the Contractor the Contract Amount of $_______**

Changes to the Contract

One thing that is a certainty on any construction project is that there will be changes to the project. In the Home Improvement Agreement, the issue of extra work and changes are specifically identified and discussed. In any other private works agreement, one of the key provisions in the contract is the one dealing with changes. The issues that arise related to this particular situation are many: who can authorize changes, what is a change, what kind of change is it, how do you make a change, how do you get paid for it, and on and on and on. A good contract will address these issues both in the prime contract and in any subcontract.

> **Note:** There is no required language for changes and extra work; pretty much anything the parties agree to will suffice. Keep in mind that if the change clause, or any provision in the contract, for that matter, is lopsided and unfair to one of the parties, a court is not likely to enforce it. Any court called on to interpret a contract will

look at both the procedural and substantive unconscionablity of any provision to decide if it can be enforced. If it can't be, because one side or the other had an unfair advantage when it was negotiated, or will get an unfair advantage if it is enforced, the clause or contract will be void and of no force and effect. So, don't think that just because a clause is in the contract that it will always be effective to accomplish what it provides.

The change or extra work provision in either the prime contract or any subcontract should identify who has authority to direct extra work and who has authority to approve changes. You do not want your field foreman signing off on change directives without the home office knowing about it. Likewise, you don't want to be taking direction from the architect if the architect has no authority to give that direction. Be clear on who can direct, and who cannot direct, changes on the project. If the person directing the change lacks authority to do so, you might not be paid!

The change and extra work provision should include a protocol for how changes to the contract are going to be made. The best way, of course, is in writing, before the changed work is begun. This is not always possible, so there should be a procedure to document the change if the work is commenced without written approval. Will the work be tracked on Time & Material, are signed daily tags required, when and how is notification provided if the work must begin without direct Owner or General authorization? Make sure it is laid out in the contract so there is no misunderstanding.

The manner in which changes are made should also be described. For example, if the Owner is directing the change, what and how is that directive made? Will it be a change directive or a request for a proposal for the work? How is the contractor to respond, and what information needs to be provided? What happens if the Owner disagrees with the costs that are provided after the contractor completes the work? If the Owner directs the contractor to proceed in a disputed work situation, what are the contractor's options, and how are those costs tracked? How and when is Notice that there is a dispute about the scope of work provided? What are the repercussions if it is not provided? These are issues that should be addressed by your Change/Extra Work provision.

Similarly, if the General directs the subcontractor to proceed with work that was omitted from the sub's original scope, how will that be tracked and how will it be paid? What documentation is required from the sub so the General can confirm that the work was done, and that it cost the amount that the sub says that it did? Is there a written directive that needs to be provided before starting the work? Who provides it? Will the sub track costs on a T & M basis for changed work, and, if so, what is required to confirm that time? Does the General's foreman or superintendent need to sign daily tags that show the workers that were on site for that work?

If the sub insists that the work is not in its scope but the General or Owner think that it is, how is that documented? Who gets the notice of the dispute, and when is it given? What information is required? Will a written directive to proceed be issued? How much notice is required for work that is disputed or changed? What are the consequences of failing to give notice? Who gets the notice? How much mark-up is allowed for changed work?

This last issue can be a real point of contention. Contractors often bid work at one rate, but then have a significant mark-up on labor and materials for change order work. Is there a maximum allowable mark-up for changed work? If so, what is it? Is the allowable mark-up in the prime contract incorporated into the subcontract? If so, say so.

These are common situations on any project, and they should be addressed in the contract. It is important to have a clear understanding about how changes will be handled, and what the responsibilities are for each party. The provision in the contract need not be long or extremely detailed, but it should be clear, and should address the issues just discussed.

There will be changes, and there will be issues related to those issues. How you deal with them is something that should be agreed to ahead of time. The Changes/Extra Work provision is one of the most litigated construction issues, so you need to be clear on what is required and how you will handle the situation when you enter the contract.

Change Orders

When you have changes or extras, you need to convert that condition into an agreement for payment for the work that is done. The Prime Contract will likely have a specified change order form that must be used. The subcontracts may have a different form than the one used in the prime contract. The contractual relationships are different, and the parties will likely express that relationship in the kind of documents that they want from each other whenever the relationship is modified, which is exactly what a change order does; it modifies the contractual relationship.

No matter how the change order is initiated or what form is used, it is important that the parties have a uniform method of instituting and agreeing to change orders, and that process, whatever it may be, should be used exclusively on the project.

Most change orders are numbered in sequence, and include a description of what additional work is being paid for by the particular change order. It is always a good idea to track added work, or disputed added work, by separate files numbered in consecutive order. When the change order is issued, you want to identify the file or files included, so that it is clear what issues are being resolved by the change order. If the entire issue is not being resolved by the change order, that should be noted on the change order. Remember, a change order is a contract, and like any contract, what is included is just as important as what is not included.

Change Order Language

Pay particular attention to the language that is provided on any change order, because it is binding and conclusive of the deal you make. Of particular importance is the scope of the change order; that is, does it include only the hard costs for the change or does it also include any impact costs; delay, inefficiency, and home office overhead? If the latter, make sure these amounts are included in the change order. If they are not included, make it clear they are not and make it equally clear you are not giving up any claim to those costs by signing the change order, no matter what the change order language provides.

It is not uncommon for change orders to include language to the effect that the party signing the change order is giving up any and all additional and related costs. It is also not uncommon, especially with Owner-issued change orders, that the contractor is prohibited from making any changes or modifications to the change order form itself, and if changes are made, the change order is invalid or void. This raises a number of thorny legal issues that will be discussed later; things like duress, ratification, and waiver, issues that can keep an attorney busy for days.

The problem can be avoided by excluding the language waiving any related costs. Either cross out the language, or send a letter that reserves your rights to assert further costs when they become known. Likewise you have the option to refuse to sign the change order. The question then becomes, is it financially possible for you to refuse to sign the change order and reserve your rights to claim interest and penalties later for wrongful withholds? It may be possible to assert another claim for the impact later, when that impact is finally realized. You have options in this sort of situation, and you need to weigh those options.

If the issue gets to court, the outcome will be dependent on the facts that gave rise to the change order in the first place, and the circumstances that the contractor was confronted with at the time he signed the change order. If you sign the change order, a court is going to presume you agreed to whatever you signed, and the court will be inclined to enforce the terms of the change order as it is written. If there is a dispute about the change order, the court may look to the facts of the particular situation to determine if it is fair to bind the parties to the agreement.

Therefore, good contemporaneous documentation of the dispute will be critical. If you sign the change order without making any effort to express your concern about entitlement to additional costs, or the language of the change order itself, you are going to be stuck with the deal you agreed to and not the one you think you deserve.

Progress Payments

Unless you are doing the work for free, you will want to get paid. Therefore, the procedure for getting paid is one of the most important aspects of any contract, be it the prime contract or the subcontract. When you get paid and what you have to do in order to get paid needs to be laid out clearly so that there is no misunderstanding and no delay in processing your payment application.

If you are the General, your subcontract should require that all subcontractors and any suppliers provide Conditional Waiver and Release and Unconditional Waiver and Release forms for Progress Payments. The contract should also require these same documents from any second tier contractors and/or any suppliers. Keep in mind that the Civil Code (section 3262) requires that specific language be used for Conditional and Unconditional progress and final lien releases, or language "substantially similar" to the statutory language in order for the Release to be effective. It is best to just copy the statutory language directly from the statute, because then there can be no question of the validity of the Release. If you use a store-bought version of the Lien Release, make sure the language is the same.

(**WARNING**: This statutory language will change as of July 2012 so be aware. The new release language is included in the Appendix at the end of the book. Use the pre-July 2012 language until that time but then be sure to switch.)

> NOTE: The Releases are different, and each serves a different purpose. The Conditional Waiver and Release Upon Progress Payment will identify how much you are requesting for the particular pay period, and will state that upon receipt of that amount (when you cash the check), you are giving up all of your lien and bond rights for that amount. You are not giving up rights for amounts you have not been paid, and you are not giving up any right you have to sue for breach of contract. All you are doing is promising that if you

get paid the amount requested then you will give up any right to ask for that amount again. The Unconditional Waiver and Release Upon Progress Payment is provided with the next payment application, and confirms that the check you received for the previous payment application was good, it was cashed, and you are giving up your lien and bond rights in the amount of those funds.

If you are the General, you want to require these Lien Releases from your subcontractors and suppliers even if the Owner is not requiring them from you. A cancelled check may be evidence that the sub was paid, but the only way the sub can waive its Mechanic Lien or bond rights is by providing these statutory releases. A cancelled check coupled with the Conditional Waiver and Release is a functional equivalent of the Unconditional Waiver and Release, but there is no functional equivalent of the Conditional Waiver and Release. Be sure to get this document; it is important.

The contract may also require that you provide invoices and payroll reports so that the Owner and/or General know that the amounts claimed on the payment application are indeed what you have obligations to pay. There should also be requirements that suppliers provide invoices and Lien Releases in order for you to get paid. This is very important, especially for Generals. A General will not know whether the supplier is being paid, or what they are being paid for, unless you get the invoices.

When you get them, review them. It is not uncommon for items to be included on your invoice that were not provided to your job. Suppliers often know before anyone else that a sub is in financial trouble, and suppliers are not shy about billing any of the sub's projects for amounts that are owed on the sub's other projects.

Creating a Subcontractor Fiduciary Relationship

When a subcontractor files bankruptcy, any attempt to recover funds that have been paid, but not applied to pay off obligations, is stayed. It is possible to petition the court to lift the stay to allow a state court action to proceed against the subcontractor-debtor, but it is unlikely the court will do that. Instead, an adversary proceeding must be filed in the Bankruptcy court and it must allege that certain of the subcontractor-debtor's obligations should not be discharged in the bankruptcy order.

There are two arguments for non-dischargeability that are typically used: 1- fraud by the debtor, and 2- that there was a fiduciary duty that the debtor breached. Both arguments depend on the successful proof of alter ego if the debtor-subcontractor is a corporation, but if this is established, or if the debtor is an individual, the debtor may be required to repay certain obligations despite the discharge of its other debts through the bankruptcy.

Fraud is a tough argument in the bankruptcy world. You must show that there was justifiable reliance on the representations of the debtor. If the debtor's progress payment included line items for supplies and wages, it might be justifiable for you to rely on that representation. It might not be justifiable if the union had previously informed you of delinquent contributions, or a supplier filed a Stop Notice.

Even more problematic is that at the time the debtor's representations are made, i.e. when the pay app is submitted, there must be no intention to perform. If the billings are submitted and the sub intended at the time to pay its bills, but thereafter did not, then the necessary showing of a "present intention" to defraud is not made. In short, fraud is difficult to prove in a bankruptcy proceeding.

Conversely, establishing that there was a fiduciary relationship is relatively easy, and can be accomplished by including a provision in the subcontract that creates that relationship. It's not a guarantee that the court will find certain debts non-dischargeable, but the threshold showing is a lot easier to meet.

A fiduciary relationship can be contractually established by including language in the subcontract to the effect that any payments that are made to the subcontractor for materials, labor and/or supplies are made in trust and that those payments shall be held in trust for the benefit of the materialmen, suppliers and laborers for that project.

The provision would read something like the following:

"All sums received by subcontractor under this agreement will be held in trust for the express use of (1) paying in full for all labor, material, equipment, and appliances furnished to subcontractor in performing this agreement, and (2) if required, making payments to labor union trust funds established under a collective bargaining agreement. No title to any payment or any part of it shall vest in subcontractor or to be used for any other purpose until subcontractor has first paid in full for all such labor, material, equipment, and appliances furnished by that date to subcontractor."

In this arrangement, the sub does not have any immediate right to the funds, but only holds them as a trustee until all of the bills are paid, and then only if there is something left after those bills are paid can it keep those funds. As a fiduciary, there is a trust relationship established, and the violation of that trust/fiduciary obligation is grounds for making that particular debt non-dischargeable in bankruptcy. If nothing else, this provision gives you another argument if you find that you have been scammed by an unprincipled sub.

Joint Checks

Joint checks are an excellent way to assure that a subcontractor's supplier is being paid. Whenever you pay by joint check, put the name of the project on the check and put some language on the check to the effect that the joint check is only to be used to pay the obligations for the such and so project. This makes it clear that you are designating and intending payment only for a specific project, and the funds cannot thereafter be designated for payment of obligations on other projects.

A supplier that receives your joint check must apply it only to your project. The supplier cannot pay off its customers oldest accounts first and then come back to your project for additional funds. No Mechanic Lien can attach to the project, and no Stop Notice is valid, if the joint check has paid off the specific project obligation.

If the supplier is providing materials on a number of your projects and you are paying it directly and not by joint check, then the supplier can pay off its aging accounts with your check. It is therefore prudent to specify what project is being paid for on the check, so that there is no confusion and so you don't end up with a Mechanic Lien or Stop Notice on a project that you believed was fully paid out.

If you are paying by joint check, you should also be receiving statutory Conditional or Unconditional progress payment lien releases from the sub and supplier for each payment. Likewise, you should be obtaining statutory Conditional and Unconditional lien releases for final payments from both the supplier and the subcontractor. The cancelled joint check can serve as proof of the payment if that is all you have, so, if nothing else, pay suppliers by joint check.

The contract does not need to provide that you can pay by joint check. You always have that right. Often, suppliers will provide a joint check agreement to the General. This is fine, but be sure to read the agreement that is provided. If you don't like it, don't sign it. You don't need an agreement at all to pay by joint check.

Other Issues with Joint Checks

It is not unusual for a sub to endorse the joint check for the other party and keep the funds. If they do, you still owe the other party and will need to get the funds back from either the sub or from the bank that took the fraudulent check. Either way you will pay again.

If one of the joint check payees will not endorse the check, there is no wrongful withhold penalty that can be assessed to you, the payor. You paid the obligation. The fact that the sub and supplier have a dispute is not your problem once you tender payment.

Monitoring Preliminary Lien Obligations

If you are not paying suppliers by joint check and/or you are not receiving statutory lien releases, there should be some trigger in your accounting system that sends up a red flag when the amount left on the sub's account is about to become insufficient if the suppliers with Pre-liens have not been paid. If you have 20-Day Preliminary Notices from a subcontractor's supplier that there will be $40,000 worth of supplies provided to the project, but you have not received any lien releases from any supplier and the amount left in the sub's account is, or is about to become, less than $40,000, then you might not want to make any further payment to the subcontractor.

At the very least, you should be contacting the subcontractor and requesting an accounting and lien releases from all suppliers. Even better, call any supplier and confirm the amounts owed for your project and ask for invoices to document the amount of supplies provided to your project only. If you don't monitor this, you may find that you will be paying the supplier with your own funds, not your sub's funds.

Pay When Paid/Pay If Paid

Provisions in a subcontract that make payment of the subcontractor contingent on payment to the General are void. If your contract provides that the sub gets paid if and only if you get paid, it is not enforceable. Pay if paid clauses are void as a matter of law. Don't include this type of provision in your contracts.

Pay when paid clauses are slightly different and might be valid, but they need to be worded so that the subcontractor's lien rights are not infringed or abrogated. A requirement that the sub wait a reasonable time for its payment, or language providing that the obligation to pay the sub arises at the time the valid payment application is submitted but the time to pay the sub arises when the Owner pays the General may be valid.

These provisions are tricky, and it is not really clear when and under what circumstances the pay when paid clause is valid. If you use the reasonable time period language, make sure you define it in some manner so that it is conditioned on an event that will occur; i.e. the conclusion of any dispute with an Owner, 90 days after the application is submitted, the occupancy of the project, etc.

> **NOTE:** Any pay when paid provision cannot interfere with the subcontractor's Mechanic Lien rights. If the provision does, and if the subcontractor is unable to record a Mechanic Lien because the payment procedure requires that it wait for payment, then that provision is going to be void. The subcontractor will always be allowed to record its Mechanic Lien, provided it is entitled to in the first place, and any provision that would prevent that right is not enforceable.

48-Hour Notice

This is an important provision, and should be included in all subcontracts. It is a warning that the sub is not performing, and that if performance is not completed, or at least commenced in some meaningful way, the General will undertake the work, or will hire someone to undertake the work, and will hold the sub accountable. The contract should be clear about how the 48-Hour Notice is provided, and what the various options will be if the problem is not cleared up. This last part is important.

If you have a 48-Hour Notice provision (it can be 72-Hour or any other reasonable time) you will want to identify what the options are if the directive in the Notice is not satisfied. Some available options are: 1- the sub can be terminated; 2- the sub's crew can be supplemented and those costs backcharged to the sub's account; 3- the sub can be assessed liquidated damages; and/or 4- any other remedy you want to employ for failure to comply with the 48-Hour Notice.

The provision should also provide that in the event of termination, the sub will not be paid until all costs are known, and then only if there is anything left over. Whatever option you want to employ for failure to satisfy the 48-Hour Notice needs to be identified in the contract. Those will be your only options, so be clear what they are.

The Effective 48-Hour Notice

The 48-Hour Notice should always be in writing, and the more specific the directive the better. What exactly does the sub need to do? Do they need to get more men on the project, do they need to have a particular task completed by a certain date, do they need to provide you with submittals or materials on site?

Be specific. It makes no sense to direct the sub to "man the job and complete the work." What exactly does that mean? If the sub brings out 2 men but needs 20, they have technically "manned the job." What does it mean to complete the work? All the work, the present phase, the present task; what? Be clear about what you want the sub to do.

A 48-Hour Notice should identify specific areas and specific results that need to be completed within a certain period of time. The sub should have the option to proceed with the work in the way that it wants, but it should also be clear that however they proceed, the required end result must be met.

The Notice should also clearly state what the ramifications for non-performance would be. If you send a 48-Hour Notice, you need to be prepared to act. If the Notice is properly sent, and assuming it contains all of the required information about what needs to be done in order to cure the problem, then if those things are not done, or if they are not done according to the Notice, there need to be consequences.

What should not happen is that the General continually sends a series of 48-Hour Notices that are never acted on. Remember the story of the boy that cried wolf. If a 48-Hour Notice is sent, and then sent again and again, the sub is arguably within its rights to ignore it. Certainly any effort to terminate a sub based on repeated 48-Hour Notices that were ignored by the sub, and not acted on by the General, could just as likely be seen as a material breach by the General as by the sub.

In short, you should have a 48-Hour Notice provision in your contract. It should be clear how it will be used, and what the consequences will be if it is ignored. If you send a 48-Hour Notice, make sure it is clear what it will take to cure it; that is, make sure it is clear what you want done.

Make sure the required action is reasonable and can be accomplished. If the Notice is sent but there is no response, make sure there are consequences. Do not send repeated Notices for the same work. Exercise your options or don't send the Notice in the first place.

Conversely, if you receive a 48-Hour Notice, respond to it. Either dispute it or act on it, but do not ignore it. Non-performance is a breach of the contract, and continued non-performance is a material breach.

Indemnity

This is the provision in the contract that most attorneys look at first when there is a dispute or lawsuit. It is the provision that Generals and subs negotiate and frequently change when they are in the process of finalizing their agreement, sometimes with unintended consequences. It is a provision that can have significant consequences, and should be clearly understood so that you are not agreeing to something that will not be in your best interests.

There are three types of indemnity provisions that might be found in a typical subcontract. Cleverly, they are referred to as Type I, Type II, and Type III indemnity. The California Supreme Court has held that these distinctions are not important; what is important are the facts and circumstances of the particular situation so that both fairness and contractual rights are protected. Nevertheless, attorneys still refer to these three Types as a kind of shorthand, and it will be helpful for you to understand what is meant. The concepts are not difficult.

> NOTE: The Type I indemnity language will not be allowed in construction contracts signed after January 1, 2013. However, contracts signed before that date with Type I and II language will still be effective and enforceable. With California's 10 year statute of repose, these provisions will continue to be litigated for some time to come.

First, what is indemnity? Indemnity is the obligation to protect another from the claims of 3rd parties who have been injured or damaged by something or someone working on your project. Third parties are usually strangers to the particular contract; i.e. they are the passer-by that gets bonked on the head by a beam or whose car is inadvertently over-sprayed on a windy day. Third parties can also be workers or other individuals on the project who get hurt by someone else working on the project and who then sue you for the injury or damage that they have suffered.

Indemnity is an important protection that is contractually based and is nearly always included in construction contracts. It should be included in your contract whether you are the General, a subcontractor, or a supplier. If it is included in the contract, then it is express indemnity. If it is not included, then what you have is implied contractual indemnity; an equitable right that you may be able to assert if you find yourself in a lawsuit.

If you have an express indemnity provision in your contract, it's important that you understand the consequences of how the indemnity language is worded, because that will determine how much protection you will get and for what acts, or in the worst case, whether you will have any protection at all. The three Types discussed in the following section are all express indemnity provisions.

Note: A recent court of appeals decision, *Crawford v Weather-Shield Windows* (13), has thrown an enormous monkey wrench into the world of indemnity. *Crawford* is being asserted by General Contractors' and Developers' attorneys as a way of getting subcontractors and suppliers to pay for the General Contractor/ Developer's legal expenses. In a nutshell, *Crawford* provides that a sub or supplier may owe a duty to defend even if it is eventually found to have done nothing wrong and no indemnity dollars are paid. If the indemnity language is correctly stated, the General/Developer can force the subs to pay its legal costs regardless of fault.

Type I Indemnity

Type I indemnity is the most comprehensive. In the Type I arrangement, the person who will provide the indemnity (the "indemnitor") is required to defend you (the 'indemnitee'), and will pay any damages that are awarded to that third party for injuries or damages that may be attributable to you.

The indemnitor in a Type I arrangement agrees to "protect and hold harmless" (this is the typical language) the indemnitee from "any and all claims, losses, costs, damages," etc. that may occur, whether the event is caused by the active or passive acts or negligence of the indemnitee and/or his agents, representatives etc. The key words to look for in the Type I indemnity provision are "active and passive negligence" and "arising out of and in connection with" the work on the project. These are especially broad terms, and they have a history in the case law that the courts will be familiar with, and that attorneys will rely on, to provide the broadest possible indemnity for the person who is to be protected. Basically, anything that happens on the project is the Type I indemnitor's responsibility.

EXAMPLE 1: Let's say you are the General Contractor building a commercial building in a downtown area. During the course of the project, someone drops a hammer off the tenth floor that falls on the front hood of a beautiful new Bentley. Naturally, the Owner of the Bentley is mad, and so he sues you for the damage and also sues the Owner of the Project. If there is a Type I indemnity in the prime contract, you will need to defend the Owner and pay for any damage. If you have a Type I in your subcontract agreement, the General will pass the obligation down to the sub whose worker dropped the hammer.

The correct term for the passing of the indemnity obligation to the appropriate indemnitor is called a *tender*. The Owner will *tender* the obligation to the General and the General *tenders* the indemnity obligation to the subcontractor. If the sub doesn't accept the tender, and there is a lawsuit that results in a judgment, the indemnitee can recover all of their costs and fees. More on this later, but if you are the sub being tendered to, it is important to understand that even if you don't think it's your problem, you can't just blow it off, especially if there is a Type I involved.

Example 2: Assume you are the General on the same high-rise. You and your sub have a Type I indemnity in your contract. Your superintendent and the foreman for the sub go up to the third floor to discuss an issue. In order to get to the place where the problem is, they have to remove the barricade around the open elevator pit, which they do. After discussing the problem they walk away from the area and don't put the barricade back. Another worker walks into the area and falls down the open pit and is seriously injured. He sues your company- who pays?

Both your superintendent and the sub's foreman removed the barricade and both forgot to put it back. In the Type I arrangement, the sub pays. The superintendent was actively negligent by doing what he did, but under the terms of the indemnity obligation, the defense and the responsibility for any damages are with the sub.

Example 3: Change the situation just slightly. Your superintendent goes up alone to look at the situation; he removes the barrier; the sub's foreman comes over and discusses the problem and then leaves; the superintendent stays there for awhile and leaves the opening unprotected. Who pays? The General does, or should. Why? Because it was the superintendent's sole negligence that caused the accident, and even though there is a Type I agreement, and even though the sub's foreman came over and was there for a while, and even though the superintendent would not have been looking at the area had it not been for the sub's issue, the superintendent was solely responsible.

In California, it is against public policy to require indemnity for the other party's sole negligence or willful misconduct. Any such provision is void as a matter of law. It is important that you specifically exclude sole negligence and willful misconduct in the indemnity language itself. If you don't, the entire indemnity provision may be void.

Type II Indemnity

Type II indemnity is less inclusive than Type I. With Type II provisions, the indemnitor must protect and hold harmless for everything except the indemnitee's active negligence, sole negligence, and willful misconduct. In the earlier hypothetical example, where the superintendent and the subcontractor's foreman both removed the barrier and both forgot to put it back, under a Type II indemnity there would not be an indemnity obligation. The superintendent actively participated in the event that caused the injury (he helped remove the barrier) so there is no indemnity required under the Type II indemnity provision.

EXAMPLE 4: Assume that the superintendent is sitting in his job trailer and the sub's foreman tells him about a problem. They agree to meet at the area in 20 minutes to go over the issues. The sub foreman goes up to the area, removes the barrier, but then walks away to get something. During that time, another worker comes along and is injured.

In this case, the superintendent is at best passively negligent. He knows that the sub's foreman is going to the area and he may even know that he will remove the barrier. What he does not know is that the foreman left the area unprotected. In this case, under a Type II arrangement, the sub would be responsible to indemnify the General.

Type III Indemnity

Finally, we have the Type III indemnity provision. This indemnity is the least protective of the three provisions for the indemnitee. If you are the person/business that will be indemnifying another person or company, this is the Type of indemnity provision you will want to include. Conversely, if you are the one that will be receiving the benefit of the indemnity provision, this is the Type of indemnity you don't want.

Typically, Type III provisions provide that if the indemnitor, the one that will be holding you harmless, is negligent, and if his work causes any or all of the harm, then and only then will the indemnity obligation be triggered and only to the extent of the percentage of liability that the indemnitor is found to be responsible for causing. In other words, if I cause the damage, I'll pay for it; but if anyone else causes some of it, then I'll only pay my share. If I am not at fault at all, then I will not pay anything.

The Ramifications of Indemnity

There are a couple of things that you need to be aware of in these situations. First, and until 2013, if there is a Type I or Type II indemnity provision, and if you are the indemnitor, you are going to be dragged into any dispute. If you or your work is alleged to have caused the damage, or if you or your work may have caused the damage, and even if it is not possible to determine what or who caused the damage, under Type I and Type II provisions you are in the mix. You will need to defend the indemnitee and you (your insurance company really) will need to contribute to any damages that are assessed. If there is more than one indemnitor involved, a typical situation in most construction situations, then you will be allocated a percentage share. If the other indemnitors are bankrupt, or worse, have no insurance, you and your insurance carrier can get stuck with the whole mess.

In the Type III arrangement, that is not necessarily the case; you are only responsible for whatever damage you caused. If you did not cause any damage, then you will not owe anything. Type III indemnity assesses a percentage amount to the indemnitor based on the amount of damage the indemnitor caused or contributed to. This is why if you are the indemnitor, you want a Type III provision, and if you are the indemnitee, you don't want it.

Until 2013, indemnity is a provision that can be, and should, be negotiated. When you see the provision in the contract, if you are a subcontractor, you should want to negotiate a more limited provision. Likewise if you are the General, you want the most indemnity that you can get. Indemnity is very important and should not be overlooked or ignored. It can have quite dramatic consequences if it is ever triggered. Insurance coverage and your insurance policy may have certain requirements and/or limitations that affect the indemnity that is or can be provided.

In short, pay attention to this provision. It is very important, and while it may never be necessary to apply it, if it is ever triggered, the provision had better do what you thought it would, or there may be some real problems that are quite costly to you as either the indemnitor or the indemnitee.

Be Careful with Changes to Indemnity Provisions

If you are the General, and you have provided a subcontract to your sub for review, be very careful of any changes the sub may make, especially to the indemnity provision. It does not take much to change a Type I or II provision to a Type III. Simply crossing out a few words can make the provision read as if it intends only to cover the subcontractor's negligence.

Example: The typical Type I or Type II indemnity provision will exclude the "sole negligence and willful misconduct of the General or Owner." By crossing out "sole" in this sentence, the subcontractor may have converted the indemnity to a Type III, or at least raised some interesting legal issues about what is intended by the provision.

You need to read any changes very carefully. If the Owner requires Type I coverage from you and your subs, and the sub has made a change to the indemnity provision, then that sub is possibly putting you in a breach situation because the Owner is not receiving the indemnity that the contract requires.

Commercial vs. Residential Indemnity Obligations

Until January 1, 2013, there is an important difference between commercial projects and residential projects with respect to indemnity. In residential construction, Type I or Type II indemnity is not allowed; it is only Type III. Subcontractors can only be required to indemnify the developer or General for the subcontractor's own negligence; a Type III indemnity.

Historical Context: Recall that indemnity requires that the indemnitor pay the indemnitee's damages, its costs; and attorney fees. It was this last part that was the killer. The shower door installer on the tract of 50 homes may have put in 100 or 150 shower doors, none of which were defective. If the plaintiff Home Owner Association sued the developer, then all the trades would get named in the litigation. The shower door sub whose whole contract for all the subdivision may have been less than $20,000, could and would be required to make a modest contribution to the damage settlement fund, but then would be whacked with a percentage portion of the attorney fees for the developer or General. Those could be absurdly high, and would often exceed the contract for the sub's work.

The Legislature changed the law to prevent that type of abuse, so that subcontractors are only responsible for claims and defects that arise out of the subcontractors own work. For contracts entered into after January 2009, the indemnity provision in a residential subcontract cannot require the sub to indemnify the General or the Developer for their own negligence or the negligence of their agents or other independent contractors.

> **NOTE:** Recall there is also the *Crawford* case to contend with now. *Crawford* deals with duty to defend issues, not indemnity. This is an important distinction and it is the new battleground. The Legislature resolved one issue but the courts have created another that seems to be contrary to the purpose of the statutory change involving indemnity. Hopefully the Legislature corrects the *Crawford* unfairness too.

Why doesn't the same limitation apply in commercial construction? (It will after January 1, 2013, but until then, it does not.) The presumption is that in the commercial context the contractors are a bit more savvy. In the commercial context, contractors negotiating a contract for a high-rise, a school, a hospital, a bridge, or any other commercial structure are presumed to be sufficiently sophisticated, and will know what they are getting into when they sign the contract. The presumption is not always true, and it is equally true that subcontractors on big subdivision projects are just as savvy and just as experienced. Often they are the same builders on the residential project as are on the commercial project. Nevertheless, the Legislature has limited indemnity on residential projects to a Type III form. The same limit does not apply in the commercial context.

> **Note:** *Implied Contractual Indemnity*: This is a form of indemnity that is implied where there is no written agreement that specifies any indemnity obligation. It is, in one sense, the purest form of indemnity because the whole basis of the obligation is fairness. If someone gets hurt, or there is property damage because of something that happens on the project, then the person or persons that caused the damage are expected to pay for that damage or injury. We will discuss this in more detail in the Litigation section, but just be aware that even if there is no express (written) indemnity provision, there is still recourse against the party that caused the damage/injury in the first place.

Insurance (see also Chapter 4 following)

Your contract must have a provision for insurance. You need to specify the amount, type, and coverage you want in any contract you enter, and you need to be clear on what you will provide as part of any contract you enter. For example, if you are the General, you will need Comprehensive General Liability (CGL) for yourself. You will also want your subs to provide CGL in specific amounts.

As the General, you will also need Workers' Compensation Insurance (if you have employees) and automobile insurance. You will want all subcontractors to have the same. You will want your insurance and your sub's insurance from an admitted carrier, and you will want to be named as an Additional Insured on any other policy; i.e. your subs and suppliers.

You will want adequate occurrence and aggregate limits; typically $1M occurrence and $2M aggregate, but perhaps more, depending on the kind of project. You will want to specify a particular type of coverage, occurrence-based as opposed to claims made, and you will want a particular type of Additional Insured endorsement (these are designated by numbers). You need to specify these requirements in the subcontract, and you need to monitor compliance so that your subcontractors and suppliers provide what they agree to provide.

These requirements may become part of your boilerplate agreement after you prepare your first contract or subcontract, so that you simply cut and paste from one revised contract to another. This is a bad idea. Do not assume that you do not need to modify your contract's insurance requirements. You will, and should, make it a point to meet with your insurance agent at least yearly to go over the coverage you have, the coverage you need, and the coverage that you are requiring from subs or any others. Insurance policies change, and insurance company requirements change. You should be sure that your policies are up to date, and that you are requiring and/or offering coverage that is still available and that is the best that is available.

Unforeseen Conditions

There are two types of unforeseen conditions. Type I is when the condition differs from what was represented in the plans. Type II is a condition that differs materially from what would be expected in the area.

Type I

Type I is the more likely condition, and is the one that will cause the most disputes. In order to prove a Type I condition, there needs to be a representation by the Owner in the contract documents that the contractor relied on, and that representation must turn out not to be accurate. The contractor's reliance will need to be reasonable as well; if it is obvious from even a cursory review that the representation is not correct, then it is not reasonable to rely on it.

Example: Assume that the soils report indicates that the soil on-site is sandy loam with some adobe type features, and that there are small rocks scattered throughout the site. The test holes are shown on the Geotech report, and indicate that the testing was fairly comprehensive; that is, there were a number of holes drilled throughout the site so that the sample taken is likely to be representative of the site conditions as a whole. Finally assume that the Geotech report indicates that the potholing was done recently, and the Geotech report also states that the report can be relied on but is not guaranteed.

The contractor gets out to the site and, low and behold, there are significant areas of hard rock and some unusually large boulders that need to be removed in order to complete the excavation.

Unforeseen Condition? Maybe; probably yes.

Recall that the Geotech qualified the report by stating it was not guaranteed. This will raise a lot of legal issues, and there will be a lot of questions that will revolve around the issue of whether it was reasonable to rely on the Geotech report, whether the contractor did anything to confirm the report and, if so, what, and whether there was an affirmative representation by the Owner in the plans that would allow the contractor to rely on the plans in the first place.

The bottom line is that the contractors, General and subcontractor, need to do their own due diligence in researching and reviewing the site. For large projects in areas where there are known hard subsurface conditions, this is something that should be done as a matter of course.

It is also critically important to read the Reports. Generals in particular need to review the Geotech Report, because it will be the General that will deal with the issue when it comes up. What the grading and excavating sub excludes in its bid proposal will also be critical if the Owner takes the position that there was no representation of subsurface conditions. The General and sub are then left to battle it out. Type I disputes always involve the issue of who assumed the risk for the subsurface condition, the General, the sub, or the Owner. Be clear on what is stated in the reports and proposals, and, if necessary, do your own investigation.

> **NOTE:** For Public Works projects the Public Contract Code, section 7104, provides that if the contractor encounters hazardous waste or conditions that differ from what is explained in the plans when the contractor is working on trenches 4' or deeper, the Owner must be notified. The Owner must then conduct an investigation and decide

whether it is a changed condition or not. If it is a changed condition, then the Owner must issue a change order. If the Owner does not find it is a changed condition and the contractor disagrees, the contractor must continue to work but retains all rights to make a claim for the work involved that is over and above what was planned for or expected at bid time. Again, document the heck out of any changed condition that is found.

Type II

The Type II changed condition occurs when you find something that is completely out of the ordinary and unexpected. It has nothing to do with any representation on the plans or in the reports; by definition it cannot.

For example, if you are excavating and come across an archeological ruin, then you have a Type II changed condition. More likely, the situation will arise where you are working in an area that is all sandy loam but then find an enormous boulder or a cave or some other condition that is not what you would expect in that area.

Unforeseen conditions are always interesting when they arise, and they create interesting legal issues for all parties. They are always factual, though, so be sure to document the condition, notify the Owner, take lots of pictures, and keep excellent records of any delays, costs, and/or inefficiencies.

Force Majeure

Force majeure is defined as a major uncontrollable change in conditions, such as war, government action, martial law situations, government overthrow, or natural catastrophe (think Katrina). You cannot do the work because nobody can do anything. You are excused from performance, and there are no consequences for your non-performance.

What happens when the catastrophic event occurs in another part of the world, or is caused by something that occurs in another part of the world? This situation occurred when there was a worldwide spike in the cost of steel. Contractors that had bid projects before the price of steel went through the roof got stuck with bids that they could not perform. Likewise, there were those contractors that bid work after the prices went sky high, but when the prices stabilized and came down, these contractors had some nice added profit.

Was the Owner obligated to issue a change order to cover the higher costs? Conversely, could the Owner ask for a credit in the situation where prices came down after the contract was let? Is this type of event covered by a force majeure clause?

It's an open question whether force majeure can be used to avoid performance or to renegotiate price when something of a more global scale, as opposed to something more local, or even national, occurs and impacts performance and price. Depending on the circumstances, it is an argument worth considering, but expect a lot of push-back.

Attorney Fees

The American Rule is that the parties in a dispute pay their own legal fees and costs unless there is a statute, law, or contract provision that provides otherwise. You cannot do much about statutes and laws, but you can do something about contracts. In this writer's view, attorney fees provisions are the absolute worst things that you can include in a contract. They are to attorneys what honey is to bees.

Therefore, in my opinion, never put them in your contract. Put a provision in your contract that specifically states that in the event of any dispute in any forum, trial, mediation, arbitration, any forum, that each side will bear their own fees and costs. If you see a fee provision included, cross it out; never agree to this provision if you do not absolutely have to for some reason. In the construction industry there are enough other ways that attorney fees can get dragged into the mix; you do not need to add any more.

Attorney fees provisions provide that the "prevailing party" will be paid their "reasonable attorney fees." Two issues here; what is the prevailing party and what is reasonable? The prevailing party in a dispute is the one that gets the result that it wants, not necessarily in degree but in substance. In other words, if you sue for $500,000 and recover $50,000, you can still be deemed the prevailing party. Your attorney fees to recover that $50,000 may dwarf the actual recovery. If you are on the losing side, this is not a good thing, because you pay your own attorney fees and you pay the other side's, too.

Reasonable fees are basically whatever the judge thinks the attorney should be paid. Some judges will look closely at the fee request; some make a cursory review at best. The fees may be completely disproportionate to the recovery, and may not be reasonable by any standard, but that will not stop the judge who wants to award them. They are totally discretionary, and the judge's discretion usually cannot be questioned on appeal.

Attorney fees are bad for other reasons, too. Fee provisions actually encourage litigation, which is precisely what all contractors should want to avoid. If an attorney sees the fee clause, he is more likely to take the case, all things being equal, than if he doesn't see one. It's an unnecessary incentive.

Fee provisions also have the unintended consequence of driving up the cost of settlement. If there is no fee clause, the parties are more likely to settle, and settle early, and save the cost of attorneys, than if one side or the other has some advantage in the case and there is a fee clause. The attorney's fees become, if not an impediment to settlement, at least a further consideration about how much it will cost to resolve the case. The more fees are expended, the higher the price goes to settle the case.

There are any number of other ways that fees are introduced into the mix. Statutory attorney fees are available for prompt payment violations, claims on bonds, and claims on Stop Notices. Why add to those by including an attorney fee provision in the contract? If you take nothing else from this book, it should be that there should not be any attorney fee clause in any contract for construction, or any other contract, for that matter. Others differ on this, and there are probably good reasons why fee provisions should be included. I've never heard them, but I'm sure they're out there.

Scope of Work, Plans and Specs, Dates

Identifying these provisions is critical to forming any complete construction contract. Likely these items are already identified as part of the bid that was provided, and they will also serve as the basis for any further negotiations that may have transpired related to the work. Make sure that what is agreed to is what is included in the contract, and, just as important, make sure that any exceptions or exclusions are noted.

As a practical matter, a General should be the one to prepare the subcontract. General Contractors should never sign a subcontractor's proposal. Never! Generals do sign them though, and when they do, the subcontractor is put in the catbird seat. It completely defeats the purpose of being the General.

Generals usually have their own subcontracts, and these can be long and complicated, and cover everything in the world, or they can be shorter and concise but still be effective and potentially very oppressive. However it is presented, subcontractors must read the contract, know what it says, and should not sign it if they can't abide by the terms, or the terms are not what were bid.

> **Note:** There are a number of form subcontracts, some good, some not so good, depending on what group is preparing the form agreement. The AGC has agreements that favor the Generals, and the AIA favors the design team. Both are good documents. They are legally binding, and can be modified and still be enforceable. Remember, contracts are whatever the parties want them to be, so whatever you agree to in the agreement, you will be bound to do. If you modify the form document in any significant way, you should have it reviewed by an attorney so that you are sure it is enforceable, and that it does what you intend and not something else.

Incorporation & Integration Clauses

Subcontractors usually demand that their Proposal be included and made part of the subcontract. This is done by incorporation language that says something to the effect that, "The proposal attached to the subcontract is made a part of the agreement by this reference." By the time the contract is finalized, the Proposal may be pretty well marked up, with lined-out parts, handwritten changes, and lots of initials by both parties. It is usually better *not* to re-write the Proposal, but instead to use the marked-up original to show there was a negotiation. The marked-up version is what is attached to and incorporated into the contract.

Most contracts also have an integration clause. This is a provision that provides that whatever is included or referenced in the contract at the time it is signed is part of the contract. If it is not identified or referenced, it is not part of the agreement. Any pre-agreement negotiations, and any documents that are not incorporated by reference, are not included in the agreement, and are, for the most part, not relevant.

The integration clause may also provide that the agreement is "fully integrated," which means that whatever is not mentioned in the document is not included. Therefore, if you see this type of clause in the agreement, make sure that you are comfortable with the contract, and that any documents you want to include in the agreement are mentioned and included by reference or attachment, otherwise they are not part of the agreement.

Any later discussions, or subsequent agreements after the contract is signed, are not covered by the integration clause. Think change orders; subsequent agreements that modify the original agreement. Any other later agreements or modifications are also not covered by the integration clause.

Inclusions By Reference

From a subcontractor's point of view, make sure that the subcontract that you sign is consistent with what you offered to do. Check the specifications and the plans, and make sure the plans that are referenced are the plans you used to prepare your bid. Watch for language in the Subcontract that says that all the work in section 7700 (for example) is "included complete, whether or not it is shown on the plans and specifications." If you have a problem with that language, cross it out or modify it. If you do not cross it out, you have agreed to do all of whatever is required in section 7700.

Subcontractors need to be careful, too, about other sections that are referenced in the particular specification called out in their subcontract. For example, if section 7700 includes a reference to other "related" sections, and if those other "related" sections call for other work related to your work, your contract may include that work, too.

For the General, watch the sub's Proposal language. A lot of times, the language is vague or conflicting. The grading sub may include sidewalks and driveways but exclude ramps. What is a ramp versus what is a driveway? Be clear, and pay attention to the particular scope that you are giving to the subcontractor. And for Generals, too, make sure that you identify the plan sheets and spec section correctly and completely. It is acceptable to reference "the plans and spec sections complete" rather than particular work, but if this is an attempted catch all phrase and sloppy drafting, and if the matter gets to court, the ambiguity in the scope is construed against the party that prepared the contract, in this case, the General.

> **NOTE:** In the prime contract between the Owner and the General, there is a similar concern, but it is not so great. The General usually has the entire scope of work, and so the understanding of what is in the bid is not really much of an issue. When there are multiple prime contracts issued on a project, usually for different trade packages, i.e. electrical, site, plumbing etc., there is a potential for overlap. Be clear what it is you are required to do as your scope of work, and confirm that the scope you intend to perform is the scope that is included in the agreement you sign.

Liquidated Damages & Actual Damages

This is an often-overlooked provision in a contract, but one that can have some significant impacts and implications for you as a contractor or subcontractor. Liquidated damages are, by definition, a dollar amount for delay damages that are agreed to before the project begins. They are agreed at the beginning of the project because the damages cannot be known with any certainty at the time the contract is signed. The parties understand that there will be damages if there is a delay; they just can't quantify them exactly, so they stipulate to them. It is not uncommon for liquidated damages to run from a few hundred to many thousands of dollars per day, every day that the project is delivered late.

There are a couple of things to keep in mind with liquidated damages ("LDs"). First, if there is an LD clause in the contract, that is the only measure of delay damages that the Owner can assess. For example, assume you are the General on a subdivision project that has a $2,500 per day LD clause. Assume the project is delayed and the actual cost of the delay to the developer is closer to $15,000 per day. An LD provision does not allow the developer to disregard the LD provision and come after you for actual damages. Therefore, a low LD amount may be a good thing.

Assume the same scenario; you are the General, but the LDs are $30,000 per day. If the actual cost of the delay to the developer is closer to $5,000 or some other significantly lesser amount, then you may argue that the LDs are not reasonable. This is a tough argument to make, because, first, you signed the contract, and second, they were an estimate in the beginning of the project before the developer could know the real cost.

However, if this developer has been building subdivisions and has had other projects delayed and has known and identified costs for delay, then maybe there is an argument that the LDs are excessive and the developer should be limited to its actual costs. Recall that the basis for any LDs is that the delay costs are not known and cannot be known. If that is not the case, the court could convert the LDs to actual damages. Of course, be sure this is what you want and that actual damages are not greater than the LDs.

Typically, LDs are assessed for delay costs. Therefore, an Owner will insist that there are other actual costs that are not covered by the LD clause. For example, the Owner will claim that the cost of penalties, or lost sales, or lost favorable interest rates, are not delay damages but are other actual damages. Maybe yes; maybe no. The more experienced Owners usually include language that differentiates delay LDs from other "actual" damages, thereby leaving open potential recovery of all delay-related costs.

When Liquidated Damages Are Available

Liquidated damages are only available to the Owner if the contractor causes the delay. They are not recoverable if there is a concurrent delay; i.e. both parties contribute to the delay event, or the delay events cannot be segregated between the Owner and General. In other words, if both parties cause the delay; the General is slow and the Owner has provided a lousy set of plans, then there is no LD assessment. The contractor and Owner do not get any money for delay, but the contractor may get a time extension.

Second, the Owner has to prove its claim that the contractor caused the delay. Just because the project finishes late does not mean that LDs are justified. If the Owner is wrong about an assessment and withholds LDs, there can be significant penalties and interest assessed under the prompt payment provisions for wrongful withholds. Of course, while all of this is being sorted out, you, the contractor, are working without being paid because the Owner is holding the LDs against your progress payment. If you are a sub, the General may be allocating a pro rata share to you as well, so basically the contractors are funding the project.

Finally, recall that in private works projects, the General Contractor does have the opportunity to negotiate the terms of the agreement. If the LD provision seems unnecessarily high, negotiate it. If it is not clear what is covered by it, discuss it. Make sure you know what the LD amount is and what it covers, because it can become very important if the project goes bad.

Delay and Inefficiency Claims

All delay is not the same, and depending on the kind of delay involved, you, the contractor, may be paid, you may only get a time extension, or you may have to pay. It is important to understand the types of delay so that you can understand the consequences.

Concurrent delay is where both parties cause the delay, and the delay-causing events are overlapping. If the Owner makes a change to the plan and, at the same time, the contractor has to tear out some work that was done incorrectly, then there is concurrent delay. If there are liquidated damages for delay, they cannot be assessed, nor can you, as the contractor, demand delay damages.

Excusable delay and compensable delay are terms sometimes used separately or in tandem. If the delay is excusable, it means that the contractor was not the cause. If it is compensable, then the contractor should be paid for it. However, it is not always true that excusable delay is compensable delay. Rain delay is a good example; you get the time on the schedule for the delay, but you won't get any money because it is non-compensable delay.

Inexcusable delay is just that; you have no excuse, and LDs or some other damage may attach, unless it was concurrent with some Owner delay. For example, if you fail to order the material and it holds up the project, you are not going to be paid for that delay unless the Owner did something at the same time that delayed the project as well. In that case, your inexcusable delay will not result in LDs. If there was no offsetting Owner delay, LDs are properly assessed.

Keep in mind, too, that there is delay and there is inefficiency, and they are not the same thing. Delay is an interference with the progress of the work that stops the progress; i.e. you are told to stop work by the Owner, and you need to wait until for the Owner to do something before you can start again. This is the classic example of delay.

Delay can also be a late answer to a Request for Information or an Owner directive that completely screws up your as-planned schedule. You are delayed because you need to stop what you are doing until you get the answer to the RFI, or until you can put the further Owner directive into play and start working again. For example, the Owner tells you that she wants another dormer in the new roof you are putting on, and you need to re-order the materials and re-engineer the whole roof structure.

Inefficiency is different. Typically, inefficiency is the consequence of repeated changes or the failure of the Owner to perform its obligations in a timely and efficient manner. You can still work, but you cannot make the type of progress you planned to make when you bid the project. For example, if the plans are bad and you constantly need to get direction from the architect, so that you are jumping around the project as you wait for the clarification rather than doing the work in an organized and systematic way, then you are inefficient. If the Owner constantly changes the plans, then you are inefficient. Inefficiency is an interference-type claim and is more likely to show up on your job cost reports rather than on your schedule, although you will see it there too.

Inefficiency is not something that can necessarily be documented on the project schedule the way delay can be. A delay claim is, by definition, a delay to the critical path. Unless the delay-causing event is to something on the critical path, it will not count as delay. However, inefficiency is not a critical path issue at all; it is a performance issue. Did you get done what you should have gotten done?

Inefficiency claims are tough to prove, but they should not be ignored. Documentation is key, and good documentation starts at the beginning of the project and continues throughout. A good and fairly standard measure of inefficiency is the number of change orders that are issued for a project. A lot of change orders strongly suggests that the project is either delayed or inefficiently constructed, or both.

To tie it all together then, LDs are the Owner's stipulated delay damages. They are only available if the contractor alone causes the delay. They have to be reasonable at the time the contract is entered, and actual damages cannot be known or knowable when the contract is entered.

Contractors can avoid LDs by showing concurrent delay, but they only get additional time, not any delay costs. Delay is determined by review of the critical path of the project. If the impact is to an activity on the critical path, then it is delay. If the delay is not to an activity on the critical path, then it is not a delay and neither LDs nor contractor delay can be assessed.

Inefficiency is not determined by looking at the critical path, and inefficiency impact costs may be available even if delay costs are not. Both Generals and subs are entitled to inefficiency costs, although it is usually the subcontractor that has the stronger claim; they are, after all, the ones who are actually performing the work, and they have the least control of the schedule of the work.

Warranty

Warranty is an important issue, and it needs to be addressed in any contract. Any subcontract warranty requirement needs to be consistent with the warranty in the Prime Contract. As with everything else in a construction contract, the General and the subcontractor need to read the warranty requirements in the specifications, and all parties need to understand what is required. They then need to confirm those warranties in the contract, if they can.

For example, if the HVAC equipment goes in, but the project doesn't complete until a year or more afterward, and if the equipment is never turned on or used, then the question will be: when does the manufacturer's warranty start to run? The Owner may think the warranty begins when the project is complete. The manufacturer may think it begins when the equipment is installed. The contractors may think that it begins when the equipment is started up and tested. Everyone has a different understanding that will affect when the warranty period begins. All the players will be using a different timeline if it is not clearly spelled out in the contract.

It is also important to assure that the required warranty periods are the same for contractor and manufacturer. Be especially aware of any extended warranties that the Owner may want. Can you provide them, and will the manufacturer give them? Likewise, be especially aware of warranty disclaimers that manufacturers provide with the equipment. It is not unusual to find disclaimers of all express warranties or disclaimers of suitability for a particular use. These disclaimers need to be carefully reviewed, especially if the Owner specifies the warranty period.

If the Owner's requirements don't line up with the manufacturer's restrictions, if there are exclusions in the manufacturer's warranty that are inconsistent with the specifications, then you, the contractor, must line them up or inform the Owner that the equipment does not come with the specified warranty. Make this an issue between the Owner and the manufacturer as much as you can, and document any issues related to warranty in writing. If you are the General or the sub providing the equipment, it is your responsibility to make sure the warranty is in place for the correct period of time, or be excused from that requirement.

Many contractors only look at the warranty requirements at the end of the project. It may be too late by then. The better time to review warranty issues is when you are signing the contract with the equipment supplier. That is your best bargaining position, because if you need to change manufacturers, or strong-arm the manufacturer, the time to do it is before the contract is written, and/or before the installation of the equipment, and not after you have paid for and installed the equipment.

Keep in mind, too that the different components of a project will often have differing warranty requirements. The Owner can also impose its own warranty requirements. The Owner's broad-based warranty requirement may not be suitable or possible for all of the items that come with warranties. HVAC equipment will be different from roofing materials. Different roofing materials come with different warranties; metal roof is not the same as built-up roofing. Whether the warranty specification can be met for each product can be problematic. Check with the manufacturer, and if the warranty can't be provided as required, advise the Owner, and the sooner the better.

Warranty is an especially important part of any contract. It is very easy for the various parties to the construction project to have different understandings of what is required, when it is required, and how it is obtained. The Prime Contract and the subcontract need to be clear, so that there is no misunderstanding.

Choice of Law

This provision is not usually an issue, but in those limited situations where it does rear its ugly little head, it can be a problem. Choice of law provisions in contracts can require you to litigate any disputes you may have under another state's laws or even in another state. Most often, you will find these types of provisions in a supplier or manufacturer's proposal, if the supplier or the manufacturer is from out of state. If you see a provision in a contract, purchase order, proposal, or anywhere else that says you will be bound by the laws of the state of X or that all disputes will be litigated under the laws of the state of X, cross it out, put in California, and initial it. California law should apply to all disputes that may arise on any project done in California.

You may also see a requirement that you use a particular venue for any litigation or dispute that you have. In most instances this is acceptable, and it is not unusual in private contracts for the parties to agree to use a certain court to resolve their disputes. If the project is in San Francisco, and the parties agree that any dispute will be resolved in San Francisco Superior Court exclusively, that's fine. What is not fine, however, is where, for example, the contract requires that all disputes will be heard in San Francisco Superior Court, but the project is in Los Angeles.

Venue is proper in a number of different locations, depending on the location of the project and who the defendants are and where they reside, but the parties cannot confer jurisdiction on a court through a contract provision. You do not have that authority. The county where the project is located is typically the venue for any litigation of a construction dispute.

> **Note:** If the contract or purchase order calls for arbitration, and the contract specifies a location for the arbitration in another state and/or with another state's laws, it is not so easy to get this changed if you don't do it upfront. Cross out the other state and put in "California". Other factors come into play, and, depending on the facts and the parties, it may be that the other state location and law will

apply, but you should try to change it if you can. Again, be careful with your contracts and read the Choice of Law provision.

Conclusion

Contracts need to be reviewed at least on a yearly basis. They need to stay fresh, and they need to be responsive to your company's project needs. If you do primarily private works contracting, but then take on a public work project, or vice-versa, you will need to review your boilerplate agreement and make sure it is responsive to the new type of project. If the provisions or any one of them, in your contract are not applicable, or worse, are confusing, then they are counterproductive and can cause more problems than they prevent. That is exactly what a construction contract should be intended to do; prevent disputes.

The clearer the agreement is about how the parties will act on the project, the better. The best way to assure that the agreement is clear is to read it and update it to account for new changes in the law and any problems you experienced on other projects. Most changes will usually be small, but that does not mean they will not have significant implications. You want the best agreement you can have, and contracts are typically a work in progress. Don't be afraid to change the language, but if you do, make you sure that you understand what the change will accomplish and that you intend it to accomplish what it does.

Chapter 4

A Brief Discussion of Insurance

First, some terminology. The general coverage that you will purchase to protect you and your company is the Comprehensive General Liability policy; CGL for short. This is what covers you for any injury or damage that you or your employees cause by doing the kind of work that you do. A CGL policy is purchased through an agent for one of the many insurance companies, the "carrier," that offer insurance. Your agent is an extremely important individual in getting you the correct policy with the correct amount of coverage, limits, and for the best price. If your company is big enough in dollar volume, you may be able to negotiate the price of the policy, but in all likelihood you will deal with the third party agent who may represent or have access to any number of insurance companies, and you will rely on your agent to negotiate the best price for the best and most coverage.

Claims-made and Occurrence-based Policies

The CGL policy provides the basic insurance coverage for you, in the event that you are sued and the claim is based on some personal injury or property damage that you (or your company) caused. Remember, your insurance will not cover you for breach of contract actions, including warranty matters. It is only when there is an injury or damage that results from your acts, or failure to act, that the insurance will come into play. There are two types of policies that can be obtained for Comprehensive General Liability purposes: claims-made policies or occurrence-based policies. Most contracts and/or subcontracts mandate that coverage is placed on an occurrence form policy.

Claims-made Policies

The claims-made policy is one that insures you for any claim that is made during the policy period. Assume you have a policy for the period January 1, 2009 to January 1, 2010. A claim is made on November 3, 2009. You will be covered. If the claim is made on January 2, 2010, you may not have coverage, depending on whether there is 'prior acts' coverage in the subsequent claims-made policy that you purchased. The simple (very simple) way of looking at this is that if you have a claims-made policy, it will provide coverage only for claims made during the time the policy is in effect.

A slightly different version of this type of policy is the claims-made and reported policy. Under this type of policy, you are covered only for claims that are made and reported to the carrier during the policy period. What is a"claim", and how it is reported, is dependent on the language in the policy, so if you purchase this type of policy, you need to review it very, very carefully.

Occurrence-based Policies & the Continuous Trigger

The occurrence-based policy provides coverage for any "occurrence" (a defined term in the policy) that occurs during the policy period. This is the type of policy that most contractors will purchase. It is the most comprehensive, and usually the most readily available. If you have this type of policy, you should keep a copy of all the yearly policy renewals that you receive, because even older policies with different carriers may have coverage obligations for claims that arise.

This is so because in California there is something called a continuous trigger that provides that if the event causing the damage is continuous, so that the damage is repeated and continuous over a period of time, then all the policies in place during the period of the continuing event have coverage obligations. If, for example, there is an underground leak that causes subsidence, which then causes damage to the structure, but the leak is not discovered for a number of years, and during that time more than one occurrence-based policy has been issued, then all the policies that are in place during the time that the "occurrence" took place are on the risk.

The "continuous trigger" is a concept that was created by the courts. The case that gave rise to this theory of continuing obligation/liability is *Montrose Chemical Corp. v Admiral Insurance Co.* The insurance industry thereafter reacted to the court's creation of the continuous trigger by including language in their policies that exclude this type of risk. If you are discussing insurance with your agent you may hear, and if you don't hear it you may want to ask about, the so-called *Montrose* exclusion that could be part of your policy. Have the agent explain what your policy will cover, and how it might apply to events that give rise to the claim if the beginning of the "occurrence" cannot be pinpointed in time.

> **NOTE:** Be aware of any 'sunset clause' in your policy. These clauses limit the time in which a claim can be made for construction defect. These are very bad clauses that you never want on the policy. Ask your agent to explain and identify any of these types of provisions in the policy.

Cost Considerations

A claims-made policy is generally less expensive than an occurrence-based policy, but the occurrence-based policy is probably a better investment for contractors. This is because the event causing the damage may not manifest immediately, and you may not even know that there is a claim. If the claims-made policy excludes events that occur under an earlier policy or before the inception of the present policy, then there may not be coverage. If the policy is an occurrence policy there will likely be coverage, because the event causing the damage will not necessarily need to be reported during the policy period in order to be covered. It may be that a particular policy from a particular carrier will not be renewed, but that policy will still cover the event if the occurrence giving rise to the damage is "continuous."

It is not advisable to jump between the two types of policies, first having a claims-made and then an occurrence-based policy. For example, if there is a claims-made policy in place, and the event causing the damage occurs during that time but is not realized, so that no claim is made, there is no coverage from that claims-made policy. If the damage is continuing and is then reported to a subsequent carrier, there is potential coverage, depending on what exclusions might be included in the subsequent occurrence policy; i.e. a *Montrose* exclusion.

One way around this dilemma is to purchase what is euphemistically called "tail" coverage. This is a continuation of the coverage under the claims-made policy for a particular time period, so that if a claim is made based on something that occurred during the claims-made policy period (but was not reported), then that claim will be covered to the extent that it would have been if it had been made during the actual claims-made insurance period.

Defense Inside or Outside the Limits of Insurance

Defense costs can be a major expense in the resolution of a claim. Some insurance policies therefore include defense costs inside the limits of insurance, so-called burning limits policies, which can erode a contractor's limits of protection in the event of a serious claim. For example, with defense inside the limits, if a contractor is sued and is forced to pay a judgment of $1,000,000 and their defense costs (attorney fees) are another $500,000, the insurer will pay the first $1,000,000 (as specified in the per-occurrence limit), and the contractor will be responsible for the rest. On the other hand, if defense costs are outside the limits, the insurer would pay those costs plus the $1,000,000 judgment, which means the entire claim is covered. Obviously, it is always preferable to have a policy where defense costs are outside the limits of insurance. Talk to your agent to be sure.

The Insurance Agent

Insurance is a fairly complicated area, and having a knowledgeable and reliable insurance agent is important for a number of reasons, not least of which being that you want to know what you are buying and why it is the best for your business. Insurance is a necessary evil, but it is more necessary than evil. Get the most you can from the best company you can. Having a really knowledgeable and reliable agent is critical. Find a good one!

Admitted and Non-Admitted Carriers

In California, insurance companies are either admitted or not admitted. You can buy a policy from a non-admitted carrier and often those rates can be less than what it would cost to get insurance from an admitted carrier. However, that reduced cost almost always is accompanied by a host of additional coverage exclusions and/or policy conditions. In addition, non-admitted insurers frequently make defense costs inside the limits of insurance – not good – whereas admitted insurers almost always leave defense costs outside the limits. Furthermore, because 'non-admitted' carriers aren't licensed in the State they are operating in, they have to sell their policies through a wholesaler (aka, an MGA – managing general agent), who will almost always charge a 'broker fee' for their services. Finally, because the policy and application goes from the carrier, through the wholesaler, through the retail agent, and then to the client, the lines of communication can often be muddied in the event of a claim.

"Admitted" carrier means that the insurance company has agreed to be bound by the insurance laws of California. It also means that these carriers have been reviewed by the California Insurance Commissioner and have been found to have enough financial wherewithal, called reserves, to be allowed to sell their products in California. In the event that the admitted carrier goes bankrupt, then the state government fund (CIGA) that is set up for insurance company bankruptcies will step in to cover at least a part of the loss. Admitted carriers contribute to this fund, and only admitted carriers can access this fund if the carrier goes out of business. The maximum coverage is only $500,000, but it is better than no coverage at all.

Insurance companies are not all the same. Some are quite good and reputable and will pay if you need to submit a claim. Others are more than just a pain to deal with; they can be scoundrels that will leave you hanging when you need their help the most. Deal with the reputable ones, and ask your agent about any carrier that may be issuing your policy. Make sure they have a good track record, and that they will be there when you need them. If you only find out how bad they are when you turn in your claim, it's too late. You will have not one but two issues to deal with, and neither will be much fun.

NOTE: You can check a company's rating by searching the AM Best database at www.Ambest.com (ratings & analysis section). The rating is comprised of 2 parts – the letter and the number. The letter (A+-F) is the grade given to the insurer. The number indicates the size of the company (VI-XV). Generally any company under consideration should have a rating of A- or better. Most construction contracts/subcontracts will require a minimum rating by the insurer (typically A- VII), and it is important to be sure that your policy is in compliance with that requirement.

Excess and Umbrella Policies

Another policy that you may wish to purchase is an excess policy. The excess policy can provide additional limits over the CGL, as well as the business auto, and employer's liability section of the workers' comp policy. The excess policy kicks in when the limits of your CGL are expended on a particular claim or in the policy period. Your primary CGL policy will have coverage that is specified for individual loss and aggregate loss. Typically it is $1 million individual and $2 million aggregate, but it can be more or less depending on what you purchase. If the loss you suffer is over the aggregate limit, there is no more insurance coverage available to you unless you have an excess policy. Excess insurance only applies when you exhaust the underlying policy limits. Excess usually covers what is covered by the primary policy, and it will "follow" that policy with respect to what is covered and not covered.

The excess policy may provide you with coverage for another $5 million, $10 million, or any other amount you may want to purchase. The rates for excess coverage are generally less than the rates for the primary CGL, so if you have the need for more insurance, an issue to take up with your agent, you should look into this type of policy. Excess carriers are also either admitted or non-admitted, but the excess carrier is typically a much bigger player in the insurance market, and so whether they are admitted or not admitted may not be of so much concern.

There is also the umbrella policy. Umbrella policies usually provide coverage for risks that are not covered by primary or excess policies, or they cover liabilities that may exceed the amount of the primary and excess policies. The limits are defined by the policy, and the terms and conditions of any umbrella policy will determine when and how the umbrella policy will come into play.

> **Note:** Most umbrella policies today are 'follow form', meaning that the coverage's (and exclusions) are consistent with the underlying CGL – meaning that there are no additional coverages provided by it. Consequently, the terms Excess Liability and Umbrella have essentially been used interchangeably in recent years.

Insurance or Incorporation

When you are purchasing insurance, you need to get the most protection for the best price, but price should be a secondary consideration. Insurance is your first line of defense to any claim that is made against you or your company. You can protect against personal liability to some extent by incorporating your business, but this may be of limited benefit if you are a small company, and you personally perform the work.

Therefore, if you have only limited funds available and the choice is between insurance and incorporating as a means to protect yourself and your company from liability, go with more insurance. (If you can do both, you should.) The insurance will provide you with a defense, and will likely cover any and all liabilities that might occur. As the court found, personal liability does not go away because you incorporate. Even if you are incorporated, if the liability exceeds the insurance you may be forced into bankruptcy, or worse, you may be personally liable. More insurance coverage or, at the very least, adequate insurance coverage, is what you want. Talk to your agent and make sure the agent knows what you do, so that the insurance will meet the need.

> **Note:** In one case, the contractor was a corporation but the company was small and the president of the company also worked on the projects. He didn't do a very good job on a driveway he installed, so the homeowners sued the corporation and sued him individually. He defended by claiming the corporate shield protected him and that only the corporation, which was out of business, could be sued. The court disagreed. The corporate shield only protects against liability for business decisions that may result in liability, not for negligent workmanship. In this case, the fellow did a poor job and was personally liable for the damage that he caused by his poor workmanship. (14)

Self-Insured Retention

Self-insured retentions, "SIR" for short, is similar, but not the same, as the deductible that you might be required to pay on your CGL policy. Unlike a deductible, you have to pay the SIR amount first before the rest of your CGL policy is triggered. For example, assume you have a $35,000 SIR on your CGL policy. If the claim amount is less than that, you pay it all and nothing comes from your insurance company. On the other hand, if the claim is $36,000, then after you pay the first $35,000, the carrier pays the last $1000. On the other hand, if you have a deductible, the carrier will usually pay the entire claim and then bill you for whatever deductible you have.

The reason for deductibles and SIRs is to keep your insurance rates down. If you don't ever need the insurance, and hopefully you don't, then the SIR or a high deductible makes sense. The savings you get from the lower premium can and should make up for some of the SIR or most of the deductible. Again, talk to your agent. The risk analysis is something that you and the agent can undertake to decide what amount of coverage you want and what amount of savings you will get by having either the deductible or the SIR.

Keep in mind, though, that if you have a very high SIR you'd better have the amount of the SIR available. Just as the insurance coverage may be claims made or occurrence, so too will the SIR. If you get whacked with 2 different claims in one year, you pay the SIR (or the deductible) on both. SIR's are often prohibited in construction contracts/subcontracts, so contractors almost always opt for a deductible, typically $1,000-$5,000.

Inclusions and Exclusions

Reading the policy is not something that you should look forward to, but it is something that you, or at least your agent, should do so that you know what you are paying for. Policies are lots and lots of fine print that is difficult to read, let alone understand. There is a section for what is covered, which is typically pretty short. Then there is the section for what is not covered, for exclusions, and for definitions, and this is where your head really starts to hurt. And that's just on the base policy; you still need to go through the endorsements and riders that are separate from, and additional to, the base policy.

Make sure you know what is being excluded, especially for your particular line of work. There will be a number of standard exclusions in nearly every CGL policy. The following is a list of some, but there will be others. You need to review them, and understand them, when you get your policy. There is nothing worse than being sued and finding out that the policy you have does not cover the liability that is alleged.

Work Product Exclusion

First, and obviously, there is an exclusion for your own work. If you do a lousy job, your insurance will not cover it. This is the work product exclusion. Generally speaking, in order for there to be coverage, your work must cause damage to another's work or to some other property, or result in personal injury. Insurance does not cover contract work that is not performed or not performed properly. It only covers damage to someone else's work that is caused by your work.

Owned Property & Builders Risk

There is the "owned property" exclusion that limits or excludes coverage for property that you, as the contractor, have "care, custody, and control" over. Your CGL policy will not cover these types of losses. However, if you are the General Contractor, you may be required to obtain Builders Risk coverage, and this type of insurance will provide coverage for damage to personal property while the work is on-going.

The obligation to provide Builders Risk coverage is important to clarify at the outset of the project, and either the Owner or the General will provide it. It is also important to know when this coverage is terminated, and who and when it terminates. When the Owner takes possession of the project, there should be no further need for Builders Risk, but this is not always the case. Be very clear, because if the claim is excluded under the CGL "owned property" exclusion and the Builders Risk policy is cancelled for some reason, there may be a hole in the coverage, and it can be a very expensive hole indeed.

Liability Assumed By Contract Exclusion

There is an exclusion for coverage of liability assumed by contract. As discussed previously, contractual issues are not covered by insurance. There is one major exception, and that is indemnity. The construction contract will likely include an indemnity provision, and this will require an assumption of liability in one form or another, but it is not excluded by this provision.

Coverage Provided By Other Insurance

Insurance policies always exclude claims that can be covered by other types of insurance. In the CGL policy, workers' comp claims, pollution claims, WRAP/OCIP claims, employers liability claims, auto claims, etc. are all excluded, because there are separate policies that can be purchased to cover those items.

Residential Exclusions

Residential construction defect is one of the most litigated areas of construction law, and is consequently an area that has been given close attention by insurers. It is critical that contractors who might engage in residential projects understand which forms of residential work are excluded under their policy before beginning a given job – and the exclusions vary tremendously between insurers. Every policy excludes residential work for OCIPs and WRAP ups, because that coverage is provided by separate policies. However, there are a host of other residential exclusions that might be included on the policy such as: single-family (custom) homes, apartments, timeshares, tracts, condo conversions, and mixed-use buildings (i.e. retail on first floor, apartments on second floor).

Miscellaneous Exclusions

Other exclusions that are fairly typical are pollution exclusions (always excluded – but can be purchased as a stand-alone policy or sometimes endorsed with a sub-limit of coverage (around $100,000) to the CGL policy) for environmental contamination in one form or another; condominium exclusions that exclude coverage for work on these types of projects; EIFS – exterior insulation finish systems – almost always excluded; subsidence – critical if you work in the dirt; and WRAP or

OCIP insurance policies where all of the contractors on a project are expected to contribute to a single project-based insurance policy. Typically, your CGL policy excludes coverage if there is an OCIP (Owner Controlled Insurance Policy) and for any damage or liability that would be covered under such an OCIP or WRAP policy.

There are other exclusions that can be included either in the policy itself or in endorsements to the policy. Among these are exclusions for employment claims, discrimination claims, and asbestos and lead claims, among others. Review all endorsements and exclusions carefully, and make sure you understand them. It might be that you want to obtain the coverage that is excluded. If so, you will need to get a 'rider' to the policy in order to obtain the particular coverage.

Commercial Auto Insurance

Many subcontracts require the contractor to carry commercial auto insurance with $1,000,000 limits of liability and coverage for hired and non-owned autos. A standard personal auto policy will generally only cover a specified vehicle and driver and are often carried with inadequate limits of insurance. This can be a major problem for a General Contractor or project Owner if there is a subcontractor whose employee gets in an accident at the jobsite since the general and owner can be included in any lawsuits arising from the accident. Consequently, it is important for the General and the Owner to be sure that subcontractors have a commercial auto policy in force including coverage for hired & non-owned autos. It is also advisable to verify that any umbrella policy extends over the commercial auto (as well as the CGL) policy, because the most severe claims tend to come in the form of auto accidents.

The Insurance Certificate

The Certificate is a summary of your policy. It lists the coverage limits, the policy number(s), the various policies you may have, the AM Best rating of company, the name and address of the agent, and certain other information. It is an important document, and it often comes with attachments, the Exclusions or Endorsements, that specify what is and what is not included in the coverage. Your policy will contain some of this information, but if you add coverage or need additional coverage for a particular project, a Certificate rather than the entire policy will be provided.

Additional Insured

The Certificate will also include, usually on the bottom left, a section called "Additional Insured." This is an important part of the coverage, and it may be an important requirement in the contract that you enter. An additional insured is just that; someone that is also covered by your policy in the event that something happens that will trigger coverage. The additional insured is covered in the same amount and to the same degree that you, the "named insured" are covered. However, and this is a big however, that is not always the case.

The insurance industry has put together any number of additional insured Certificates, and each of them has its own identifying number that serves as a way for the carrier, the carrier's attorney, and/or your agent and attorney to know exactly what the limitations are for the specific Additional Insured endorsement. Some "AI's" are quite limited and don't do a heck of a lot. Others are broader and cover more. You need to know what is, and is not, included with the particular AI you obtain or receive.

If you are the General or the Owner, and you want the subs to cover you as an additional insured on their policies, it is critically important that you specify in your contract what AI Endorsement you will accept by the particular policy number. You will not have this information, (because these types of endorsements change with some frequency) so you must contact your agent and find out which is the best AI for your purposes. The agent should know, and if they don't, get another agent.

If you are the sub that needs to provide an AI, they are not hard to come by. Call your insurance agent and they will provide what you need. Typically the General and the project are identified by name as AI's on the Certificate. If you change carriers in the middle of a project, it is important to get a new AI from the new carrier so that you have coverage for the remainder of the project. This is because the requirement to name the Owner or the General as an AI is a contract requirement, and if you don't do it, you are in breach of the contract and may also be personally responsible for the coverage that the policy would have provided.

If you are the General requiring your subcontractors to name you as an Additional Insured, you will want the sub to provide you with a copy of its Certificate that shows you and/or the project named as the AI. The AI endorsement should come with the Certificate. When it does, check to make sure the endorsement is the type that you required in your contract. If you are unsure of what you are getting, call your agent and find out.

Always keep these Certificates and the AI endorsement for all subcontractors, and suppliers that may provide them, in a separate binder. These are documents that will have an importance that will continue even after the project is complete. You need to have them in case there is ever any problem on the project, so make sure you know where they are.

> **Note:** One point that is often overlooked but that can be a major problem is when you start off in business as a sole proprietorship, but then decide to incorporate your business. The insurance that you had as a sole proprietor will not cover the new corporate entity. When you begin work as a corporation and stop work as a sole proprietor, the insurance that you had as a sole proprietor stops as well. Unless you, as a sole proprietorship, are a named additional insured on the corporation policy, i.e. as an individual *dba*, your sole proprietorship, you may be out of luck and have no insurance available to cover this claim. In other words, you pay out of your own pocket. Talk to your agent when you convert from one form of entity to another, and make sure there is coverage that follows for any work you did before the change of entity form.

Additional insured certifications are important and need to be included in your contracts. They don't cost much, if anything at all, so it's really just a call to your agent. Make sure you know the type of AI you want, and make sure you get it.

> NOTE: An interesting but troubling case that came down in early 2010, *Forecast Homes v Steadfast Insurance*, threw a monkey wrench into the whole Additional Insured discussion. According to the Court of Appeals, if the insurance policy at issue has a self-insured retention requirement, and if the policy specifies who can pay that SIR, then the AI coverage is not triggered until and unless that person actually pays the SIR. If the policy specifies that only a certain person can pay the SIR, no one else can pay it! Therefore, if the named insured is bankrupt, dead, disappears, or won't or can't pay, the AI part of the policy is not enforceable, because the underlying CGL has not been triggered because the SIR has not been paid! A strange decision, and one that it will play havoc with AI's, especially when insurance companies insert those types of requirements into the named insured's SIR provision.

Waiver of Subrogation

Waivers of subrogation are a promise by the insurer that they will not take legal action against a specified entity. A waiver can be endorsed onto a CGL, auto, or workers' compensation policy, and it is often required in contract/subcontract agreements. The endorsement can be necessary because when a subcontractor signs a contract stating that they (the subcontractor) will indemnify (hold harmless) a General Contractor or a project owner, the insurer does not necessarily have to conform to the indemnity provision in the contract.

For example, if a General Contractor creates unsafe working conditions, and a subcontractor's employee gets injured on the job, the sub's workers' comp policy will pay the initial claim. After that, the insurer can subrogate (sue) against the General to recoup the costs of the claim even though the subcontractor agreed to indemnify the General. By having a waiver of subrogation in favor of the General endorsed to the policy, the sub's insurer agrees that they will not try to recover from the General in the event of a claim. Since the waiver of subrogation limits the insurer's chances of recovering on a claim, there is often a charge to add this endorsement to a given policy. The waiver of subrogation endorsement is required in most construction contracts/subcontracts, so it is important to understand the purpose of this endorsement for any contractor looking to enter into a contract agreement.

Filling Out the Policy Application

When you are applying for insurance, you will need to fill out an application. Do NOT lie, misrepresent, exaggerate and/or forget any pertinent information! If you do and the carrier finds out, your policy is voidable and can be unenforceable.

This typically comes up if you make a claim, and the carrier investigates the claim. If the carrier finds out that you misstated some important fact, like the fact you have been sued 17 times before and you said your record was clean, they will rescind the policy and you will not have insurance at all. Talk over any issues with your agent, and do what you can to get the insurance that you need, but do not misrepresent the facts. If you do, then at the time you need it most, you may not have it at all.

Workers' Compensation Insurance

Workers' compensation information needs to be provided, and it should be an included requirement in all subcontracts. If you are the General, you want to know who the Workers' Comp carrier is for your subcontractors. If they have employees, they need this insurance. All contractors with employees need this insurance.

If you are a contractor and have employees, but do not obtain Workers' comp, there can be severe consequences. For example, if you have employees and one of them gets hurt, the injured employee's exclusive remedy against you, the employer, is Workers' Compensation. They may be able to sue other parties, but they cannot sue you. If you do not have Workers' Compensation, you become fair game because there is no Workers' Comp remedy available to them. You (or your company) get to pay any damages that may be assessed, and these can be very big dollars.

Compounding that problem, in a really dramatic way, is the fact that if you don't have Workers' Compensation and you have employees, you are not compliant with the licensing law. This could result in a finding that you are not licensed, and, as you know by now, if you are not licensed during the entire time of the work, you do not get paid. Worse yet, you need to refund all the money you have been paid.

Finally, after January 2011, various enforcement agencies can issue Stop Work orders on projects where a contractor has no Workers' Comp. The project can be shut down, or the non-compliant contractor can be ordered off the job. Workers' Comp is an important insurance requirement. Do not overlook it, and do not operate without it, unless you don't have any employees. Make sure your subcontractors have it, and make sure the information is provided in any subcontract that you issue.

Chapter 5

Private Works

In private works contracting, Generals and subcontractors have some fairly strong rights that they can use against a private Owner; rights that don't exist in the public works arena. For one thing, a private works contractor can force the sale of the property they are working on to pay outstanding bills by recording and then foreclosing on a mechanic lien. A private works contractor can also stop work under the right circumstances. Coupled with the fact that private works contractors also have Stop Notice rights, and, in some situations, bond rights, the private works arena is seemingly more contractor-friendly than public works.

This is not to say the private Owner is less formidable than the public Owner. They are not. A private Owner has important rights and significant remedies, and so just as in Public Works, it is important to know what the rules are and how they apply. Not following the rules, or not knowing the rules, can be just as costly whatever kind of project you are building, so you need to be aware and do it right. If you don't, it can cost you very dearly.

Preliminary Notice

In private works, subcontractors and suppliers must provide the Owner, the General Contractor, and the construction lender (if there is one) with a 20-day Preliminary Lien Notice. This is an important first step and many of your remedies for non-payment will depend on whether you send this Notice, as well as when you send it. Make sure that you send it, and make sure you send it before you begin any work.

The purpose of the Preliminary Lien Notice is the same in public and private works; it tells the Owner, General, and lender that you will be working on the project and that you expect to get paid. The 20-Day Notice is simply a way of letting everyone that matters know that you are on the job. If you have a direct contract with the Owner you don't need to send one, but one should go to any construction lender. Workers do not need to send Preliminary Notices; their rights are already secured if they work on the project, provided of course that they can prove they worked on the project.

The 20-Day Preliminary Notice should be sent before any work is commenced on the project, but it is not fatal to the cause if it is not. A late 20-Day Notice will secure the value of the work done from the time 20 days before the Notice is actually sent, and thereafter until the work is completed. Even if the Owner knows you are on site because he/she sees you and knows who you are, you must provide the 20-Day Notice in order to perfect your rights. Get in the habit of sending the Notice as soon as possible after you get the contract and, by all means, before you start work.

Required Notice

The 20-Day Notice must include the following statutorily required warning to the Owner regarding the consequences of a Mechanic Lien:

NOTICE TO PROPERTY OWNER

If bills are not paid in full for the labor, services, equipment, or materials furnished or to be furnished, a mechanic's lien leading to the loss, through court foreclosure proceedings, of all or part of your property being so improved may be placed against the property even though you have paid your contractor in full. You may wish to protect yourself against this consequence by (1) requiring your contractor to furnish a signed release by the person or firm giving you this notice before making payment to your contractor, or (2) any other method or device that is appropriate under the circumstances.

Other than residential homeowners of dwellings containing fewer than five units, private project owners must notify the original contractor and any lien claimant who has provided the owner with a preliminary 20-day lien notice in accordance with Section 3097 of the Civil Code that a notice of completion or notice of cessation has been recorded within 10 days of its recordation. Notice shall be by registered mail, certified mail, or first-class mail, evidenced by a certificate of mailing. Failure to notify will extend the deadlines to record a lien.

If this language is not included, the Notice is not valid. The 20-Day Notice is a standard form that can be purchased at most stationary stores, but be sure that the form that you use has the required and correct language so that it is legally enforceable. Ask your local contractors group or Builders Association and make sure that it has the required Mechanic Lien language.

Registered or Certified Mail Only

The 20-Day Notice must be sent by registered or certified mail. You want to send it this way so that you have the receipt that shows that you sent the notice on the date that the registered mail says that you did. Keep this receipt and the one that comes back and confirms the delivery. Even if the return receipt is not received, the law presumes receipt in a reasonable time after deposit in the mail. Therefore, keep the receipt you get from the post office or whatever mail service you use. Staple it to the copy of the 20-Day Preliminary Notice that you will keep in your project file. Do not lose it; scan it into your computer so you can print it out if you need it.

Where to Send Notice

The 20-Day Notice is sent to the Owner, the General, and the construction lender. If you are a sub, the General is already known. The General should also provide you with the names of the Owner and any bank or other lender, if there is one. If the General does not give you this information, then the Building Department that issued the building permit will have the information, and the information should be contained on the permit. You are entitled to rely on that information, so take the names and addresses down and send the Notice.

Notice Must Be Accurate

The information contained on the 20-Day Notice is what you would expect; your name or company name, and the address, the name of the project, the type of work, in general terms, that you will perform, and the approximate value of that work. The estimated value of the work that you give should be reasonably accurate, not grossly understated or overstated. If you put down that you will be performing $10,000 worth of work but end up doing $150,000 there could be a problem when it comes time to try and perfect your rights. The Owner and construction lender rely on this information to calculate necessary funding and cash flow. If the information provided is completely unreliable and the sub or supplier claims more money than was anticipated by the 20-Day Notice, the courts, in some instances, have ruled that the Notice is invalid (void) so that the contractor had no rights.

The 20-Day Notice is important to secure rights for Mechanic Liens, Stop Notices, Bond claims, and other rights as well. If you don't get the Preliminary Notice correct, or if you don't do it at all, there can be some serious repercussions. You won't lose all your rights, but you will have a lot less of them to assert. Again, the 20-Day Notice is not a difficult document to prepare, so do it, and do it right.

Mechanic Liens

Mechanic Lien rights are an important and constitutionally enumerated right that contractors have in private works, but that do not exist in public works. It is specifically addressed to contractors (mechanics) in the State Constitution (Cal. Const. Art. 14, Section 3) and provides the right to place a lien on a private work of improvement to secure payment for the work that is performed. The lien attaches to the work of improvement and, to a degree, on the land that the work sits on.

> NOTE: You cannot record a Mechanic Lien against a public works project even if a private group will use the publicly funded project. If it is public funds and a public building, there is no Mechanic Lien right at all. (22)

Once a Mechanic Lien attaches, and if a contractor remains unpaid, then the court can order the property sold to pay the outstanding payment obligation. This sounds pretty impressive, but as we shall see, it is not always possible or practical to go through the process. It is only when the contractor is entitled to be paid for the work that a court will actually order such a remedy. It is nevertheless an important option that contractors have and should always consider if they are not paid.

Who Has Mechanic Lien Rights

Mechanic's Liens are available to any of the statutorily identified categories of "mechanics," and only to them. The Mechanic Lien laws cover the General Contractor, subcontractor, supplier, materialmen, and laborers that contribute to the work of improvement. The lien right does not extend to the suppliers of the supplier; for example, it will not extend to the supplier of steel to a fabrication subcontractor who turns the raw steel into specific beams and columns for use in the project.

The mechanic lien right also requires that the contractor who will assert the right be properly licensed at all times during the performance of the work. This is critically important; you must be properly licensed. If you are not properly licensed at all times when you are performing work on the project, you have no right to be paid at all, and you have no Mechanic Lien rights. Properly licensed means that you have the requisite license, not simply that you have any contractor's license. (Licensing requirements were discussed in Chapter 1.)

Individual workers also have Mechanic Lien rights for the labor that they contribute to the project. If the worker is not being paid, the worker can record a Mechanic Lien against the property, and the worker can then perfect that right in the same way that any contractor can. If the worker is part of a company that supplies laborers as its business, and if that company is not being paid for those workers, then the company has lien rights. However, if the company is only a referral service and does nothing more than refer out workers, there is no Mechanic Lien right for that company.

Suppliers of things such as scaffolding also have Mechanic Lien rights, even though, technically, the scaffold is not "consumed" or made part of the project. It is necessary to the project, and therefore under the law, it is used in the work of improvement, and so the Mechanic Lien right is available.

Likewise, a trucking company that is hired by a "representative of the Owner" has lien rights. The representative of the Owner is anyone who has a direct contractual relationship with the Owner, or with someone that has such a relationship. So, for example, if the grading subcontractor orders baserock from the local quarry for the parking areas, the trucking company that hauls the rock has a Mechanic Lien right, as does the quarry. But if the baserock is quarried at a distant location and is sold out of a local yard, the distant quarry does not have any lien right, although the local yard will have such a right.

It is usually clear who has lien rights on a project, but there are some unusual situations where the question of the Mechanic Lien right is not so clear. For the most part, if you are working on the project you have a Mechanic Lien right.

Recording the Mechanic Lien

If there is an issue with payment on the project, especially if the project is complete or nearly complete, the decision to record a Mechanic Lien is at hand. The Mechanic Lien needs to be recorded at the County Recorder's office, or wherever such documents are recorded. The Mechanic Lien itself is a form document that can be purchased at most local stationary stores, builder's associations, or online.

The Mechanic Lien is recorded in the County where the property is located; don't record it anywhere else or it will not be valid. There are specific time limits to record the lien, but these time limits are different, depending on who is recording the lien and what the Owner has done with respect to filing a Notice of Completion. It is also important that you are complete with your work. If you are not done, the Mechanic Lien is not valid.

Time Limits to File the Lien

If you are the General, or if you have a direct contract with the Owner, and if the Owner has recorded a Notice of Completion or a Notice of Cessation on the project, and if you are informed by the Owner that there is such a Notice recorded, then you must record your Mechanic Lien within 60 days of the Notice of Completion or Cessation.

If the Owner does not let you know that it has recorded either a Notice of Completion or Notice of Cessation, then you have 90 days to record your Mechanic Lien from the date of the recording of either Notice. You need to track these dates, and yes, it is confusing.

However, if the Owner does not record either a Notice of Completion or Cessation, time begins to run from "completion". The definition of completion is important. Completion means actual completion. Com pletion is defined as 1, the use or occupancy of the project by the Owner or the Owner's agents or representatives; 2, acceptance of the work by the Owner or the Owner's representative/agent; or 3, cessation of work for 30 days if the Owner records a Notice of Cessation or, if no Notice of Cessation is recorded, then cessation of work for 60 days. The cessation of work is for a continuous period of days.

Therefore, if there is no Notice of Cessation recorded and there is a cessation of work for 60 continuous days, then there is completion. If there was no Notice of Cessation recorded or you were not informed of the Notice of Cessation or Notice of Completion, you have 90 days to record the Mechanic Lien from completion. Therefore you have 150 days (60 + 90) to record the Lien from the date that work stopped and the 60-day cessation period begins and ends.

It is slightly different for the subcontractor that wants to record a Mechanic Lien. If there is a Notice of Completion or Notice of Cessation filed, the sub has only 30 days from the date of the Notice of Completion to record a Mechanic Lien. If there is no Notice of Completion or Notice of Cessation recorded, or you are not notified of the recording of either Notice, then the subcontractor has the same time limit to record the Mechanic Lien as the General or any Prime contractor.

There is an outline of these timeframes at the end of the book in the Appendix.

Notice of Completion from the Owner

How do you know that the Owner has recorded a Notice of Completion? If you sent a proper 20-Day Preliminary Notice, the Owner is required to notify you within 10 days of their recording of the Notice of Completion. If the Owner does not do so, then the Owner cannot argue that your Mechanic Lien is late or untimely; another good reason to keep a copy of the registered or certified mail receipt of your 20-Day Notice.

If you record your Preliminary Notice at the Recorders office, then the Recorder should send you notice when the Owner records any Notice of Completion or Notice of Cessation.

Required Notice of Lien to the Owner

Starting in January 2011, before you record a Mechanics Lien on a property, you must send a Notice to the Owner, or the person/entity that you believe is the Owner, informing them that the Mechanics Lien is about to be recorded if no payment is received. The notice to the Owner must be sent by registered mail or certified mail. If you do not send this Notice to the Owner, the Mechanic's Lien is unenforceable.

The Notice must also contain the following language in at least 10 point type:

NOTICE OF MECHANIC'S LIEN ATTENTION!

Upon the recording of the enclosed MECHANIC'S LIEN with the county recorder's office of the county where the property is located, your property is subject to the filing of a legal action seeking a court-ordered foreclosure sale of the real property on which the lien has been recorded. That legal action must be filed with the court no later than 90 days after the date the mechanic's lien is recorded.

The party identified in the mechanic's lien may have provided labor or materials for improvements to your property and may not have been paid for these items. You are receiving this notice because it is a required step in filing a mechanic's lien foreclosure action against your property. The foreclosure action will seek a sale of your property in order to pay for unpaid labor, materials, or improvements provided to your property. This may affect your ability to borrow against, refinance, or sell the property until the mechanic's lien is released.

BECAUSE THE LIEN AFFECTS YOUR PROPERTY, YOU MAY WISH TO SPEAK WITH YOUR CONTRACTOR IMMEDIATELY, OR CONTACT AN ATTORNEY, OR FOR MORE INFORMATION ON MECHANIC'S LIENS GO TO THE CONTRACTORS' STATE LICENSE BOARD WEB SITE AT www.cslb.ca.gov.

Time Limits to File the Foreclosure Lawsuit

Once the Mechanic Lien is recorded, you will need to bring a lawsuit (foreclosure) within 90 days of the date of recording. It is not clear what happens if the Mechanic's Lien foreclosure action is not brought within 90 days, and the time to record a lien has still not run out.

For example, assume the Owner has not filed a Notice of Completion or a Notice of Cessation, so that the time to record the Mechanic Lien is 150 days from the last day any work was performed on the project. Assume the contractor records the Mechanic Lien on day 35 after the work has stopped. 90 days after that, the contractor should file the lawsuit, but they don't. Instead, the contractor removes the first lien, records another one, and then files a lawsuit 80 days later. Is the lawsuit valid?

Some courts say yes; some say no. The better reasoning is that it is valid because the Mechanic Lien right is constitutional and the contractor has perfected its rights within the statutory time. Not all courts agree and there could be a legal issue that is raised under these circumstances. Be aware then that the time limit can create a lot of issues and problems for the Mechanic Lien claimant. Of course, if the Mechanic Lien action is not brought, or the Lien is foreclosed after the last possible date, then the cause of action is going to be dismissed as untimely.

Special Situations Involving Mechanic Liens

Problems arise when, for example, there are multiple units or multiple contracts for the project. If the project is a subdivision of 100 homes, there will be different issues raised than if the project is a 100-unit condominium project. Similarly, if there are different contracts for different parts of the project, there will be different Mechanic Lien implications.

If the project is a subdivision of 100 homes, then as each home is completed, the Mechanic Lien right attaches. The 20-Day Preliminary Notice is also impacted. In these circumstances, subcontractors, especially, have to be very careful. The 20-Day Notice should identify the project as a whole, if it is a lump sum contract, but if there are different models and types for the houses, and if it is possible to distinguish the cost for each, then providing a breakdown of that cost/model/and address on the Preliminary Notice might come in handy in any litigation later on.

Likewise if the project is being sold as it is built and the individual homes are each getting a Notice of Completion, then the sub or General/Prime contractor should be recording liens against the individual properties. The contractors need to allocate the amount of any lien or unpaid amount to a particular property, as well. These liens should show up in the chain of title for the purchasing Owner (assuming the contractors don't wait to record them until the last minute) so that the new Owner will be aware that the home they are purchasing is coming with a lien attached that they may be required to pay off. This creates an incentive for the developer to either pay the contractors or else obtain a Mechanic Lien Release Bond. In any event, the unpaid subcontractor's right to be paid is protected.

If the project is a condominium or townhome arrangement, the scenario is a bit different. These types of projects are a hybrid of individual Ownership and community Ownership. Nevertheless, under the law, the completion of the entire building is what constitutes completion of the project. If the project is a cluster of condominiums or townhomes, and each building contains multiple units, then the completion of the entire structure, not the individual unit, is what triggers the time for recording the Mechanic Lien.

Keep in mind, too, that the same time limits apply to single-family homes, condos, and townhomes that are constructed as subdivisions, or PUDs, as would apply to any other home or commercial structure. That is, if the Owner does not record the Notice of Completion or Notice of Cessation, and/or work stops on the project for the requisite time, the time to record Mechanics Liens is the same 30, 60, 90, or 150 days, depending on the circumstance. Don't be confused by the time to record the lien and the time to foreclose (file the lawsuit) to perfect your lien. You always foreclose the lien within 90 days of recording it, regardless of when you actually recorded the lien.

Another situation that can arise is when the grading and underground work occur under one contract and the construction of the buildings occurs under another. If there are two separate contracts, the Mechanic Lien recording requirement is separate as well. A grader cannot argue that the completion of the entire project is what triggers his Mechanic Lien rights. Those rights are triggered by the completion of the work that the grader does under the separate contract; not the completion of the entire project. Of course, if the project is small and the grading is integral with all the other work, and there is only one contract, the completion of the entire project is what triggers the recording of a Mechanic Lien.

The grader may want to pro-rate the value of the work to individual units, if this can be done. This may not be possible or practical, especially if the grading was only for streets and underground and does not include lot grading. In this instance, there may be an issue that the individual Owners can raise about the availability of the Mechanic Lien attaching to their individual property. The argument in response is that the individual units could not be completed or used without the roads and underground, so it is only fair to pro-rate costs to the individual units.

The Parties to the Mechanic Lien Action

After you have recorded the Mechanic Lien, you must "foreclose" on the lien within 90 days. This means you have to file a lawsuit to 'perfect' your rights to recover the funds that you are owed. Effectively, what you are asking the court to do is to force the Owner to sell the property so that you can be paid. When you file your lawsuit, you must name all of the lienholders on the property as defendants in the lawsuit. That means any person or entity (bank, mortgage lender, Second Deed of Trust holders, private lenders, etc.) with any interest in the property must be named in the Complaint. You get this information by having a Title Company run a Property Profile on the specific property. Anything that is recorded against the property and anyone with an interest in the property should show up on the Title Report.

If you are not certain from the Property Profile if all of the possible lienholders are listed, or if there is some ambiguity in the Profile, you can purchase a "Mechanics Lien Guaranty" from the Title Company.

Basically, this identifies anyone who will be impacted by the foreclosure action. Remember, if all indispensable lienholders are not notified of the foreclosure action, and if the foreclosure actually happens and the property is sold, it can be challenged by the un-notified lienholder and the entire process can be overturned or, at the very least, will need to be redone. Therefore, it is important to get notice to all lienholders.

Once you file the lawsuit, you must record a Lis Pendens in the County Recorder's Office against the specific property. The Lis Pendens is a notice that there is a problem with this property, and the property is at issue in a lawsuit. Any potential buyers will be on notice, and any other lenders that may be thinking of making a loan will be on notice. No one will purchase the property, but if they do, they purchase it with full knowledge of the lawsuit, so that your claims are not extinguished. It is unlikely that anyone can purchase the property with the Lis Pendens attached.

> NOTE: If you are the General Contractor, one thing to keep in mind is that if a sub records a Mechanic Lien against a residential property and then forecloses, you, the General, are obligated to defend the Owner. If you do not, and assuming you don't have your own problems with the Owner, then the Owner can come after you for costs and fees associated with the defense of the sub's Mechanic Lien litigation. If the sub was not paid because you and the sub had some other issues, and the sub initiates the foreclosure action, then you, the General, are responsible for all the Owner's damages. These damages can include any loss of favorable interest rates, lost sale of the property, or other actions the Owner intended but could not accomplish because of the Lis Pendens and Mechanic Lien foreclosure action.

Priority of Liens

Let's assume that you have properly sent a 20-Day Notice and recorded your Mechanic Lien, and you are now bringing an action to foreclose on that lien. Assume that this is the only cause of action that is alleged in the lawsuit, something that is not likely but for our purposes will help you understand what filing an action will do.

First, there is the issue of priority. If the construction lender has recorded its lien before you recorded your Preliminary Notice, then it will have priority over your lien. This means that if the property is actually sold, the amount the construction lender has put into the project will be paid off first. The remaining amount will be used to pay your lien amount and any other liens that are in place at the time. If the project is upside down financially, there will not be much, if anything, left for you to get when the project is finally sold. If your lien is one of many, then you may get some pro rata share of the recovery if the property is sold.

The Owner's Response

Let's assume that you are the only contractor with a Mechanic Lien and you bring your foreclosure action. Assume as well that either you had already stopped work on the project (a topic that will be discussed shortly), or that you finished the project, but now the Owner has decided not to pay the change orders that he agreed to earlier.

You foreclose on the Mechanic Lien, thinking you can get the court to either force the Owner to pay you or force the sale of the property, and you can get paid. It won't happen that way. In all likelihood, the Owner will file a cross complaint against you. The cross complaint will allege that you breached the contract in any number of ways, and that you also did defective work.

We will discuss the litigation process in a later section, but, for now, think insurance. Remember, litigation is a very expensive process, and the Owner has your money to carry on the fight against you (at least you hope he does). Your money will be used to fund the Owner's litigation, for the time being anyway, and depending on how much is at stake, the fight could be long and drawn out. If the Owner has alleged defective construction issues against you, and has alleged damages as a consequence of that defective construction, you need to notify your insurance company and tender your defense to the carrier. This is a fairly straightforward process, especially if you have a good insurance agent. When you receive the cross complaint, notify your agent, or, if you have an attorney, send the cross complaint to the attorney and he/she will tender the cross complaint to the carrier and demand a defense.

If the tender is accepted, the entire litigation should be defended and prosecuted by the carrier. The insurer will appoint an attorney to defend you in the case, and the carrier will pay the attorney fees for that defense. The carrier will also prosecute your Mechanic Lien issue as a possible offset to any claim by the Owner, but you will need to have your own attorney to assure that your rights and interests to any funds are fully protected. You don't want to negotiate away the unpaid amounts as offsets to the claims of the Owner. You may have some attorney costs, but the insurance company should pay the majority of litigation costs and this will even out the playing field with the Owner, at least to some extent.

Valuing the Mechanic Lien Action

The biggest question/decision that you will confront when considering any lawsuit, and especially a Mechanic Lien action, is, is it worth it? If the amount owed is small, less than $7,500, consider a small claims action. It's cheap and it's fast. If the amount is more than $100,000, it's probably worth going forward with the case, although this is still a close call.

It is important to engage an attorney at the outset of any anticipated litigation to help you review your options. It's rare for a Mechanic Lien case, or any construction case, to end up in trial, but they can and do, so understanding your options and properly evaluating the likely outcome is important before you file the action. The money spent pursuing the Lien amount can, and often does, dwarf the Lien amount itself.

Mechanic Liens are important legal rights, but they are only established through litigation, and litigation is very expensive. If you are not being paid on a private works contract, then by all means include a Mechanic Lien action in your lawsuit, but the amount at issue needs to justify the cost of the litigation that will be required. The Mechanic Lien claim is a very important one, with good strategic and tactical advantages, and it should be exploited to every advantage. It has its downside though- attorney fees- so it needs to be evaluated completely and carefully.

Stop Notices

A Stop Notice, in either a public or private project, is a notice to the Owner or construction lender that tells the Owner or lender that you are not being paid for your work on the project, and that you want the Owner or the lender to hold back a certain amount of money so that you can be paid. A General Contractor may not serve a Stop Notice on the Owner, but subcontractors and suppliers can. The General and the subs/suppliers can all serve a Stop Notice on the construction lender. The Stop Notice is a claim on the available funds for the construction, and it is an important option, although not without its downsides and potential for abuse.

The only meaningful way to file a Stop Notice in private works is to file a bonded Stop Notice. If you file a bonded Stop Notice, a 20-Day Preliminary Notice must have been served on all required parties. If you file a bonded Stop Notice without having a 20-day Preliminary Notice in place, the Stop Notice is not valid. Therefore the 20-Day Notice is absolutely necessary, as in it is a prerequisite, so make sure you have the 20-Day Notice in place!

Unlike Mechanic Liens, the Stop Notice can be served at any time during the course of the project. The project does not need to be completed before serving the Stop Notice. If you are not being paid, it is not advisable to wait until the end of the project or, for that matter, until your work is completed, before sending the Stop Notice. It may be too late by that time. As we will discuss, when it is clear that you are not being paid, or that you are not going to be paid, you need to file the bonded Stop Notice.

You do not record the Stop Notice; you serve it by mail or by hand delivery on the Owner and/or lender, and the General. It is not absolutely necessary that the General get served; the General will be notified by the Owner or the Lender in any case. If you are the General, though, you want a copy of any Stop Notice for your records.

Some Considerations Before Filing the Stop Notice

If you are a sub and you serve a bonded Stop Notice, the first thing that you are guaranteeing is that you will not be paid by the General. The Stop Notice is, after all, telling the Owner or lender not to pay the General because the General is not paying you. If the General was not paying you before the Stop Notice, then in all likelihood the General will not pay you after the Stop Notice is served. Presumably there is a reason why the General was not paying you in the first place, and hopefully it is a good reason, because the General and the sub are now in a dispute situation.

When the Owner or lender gets a Stop Notice, whether bonded or not, they will likely withhold the funds. This is because there is also a Mechanic Lien right, and the Owner will not want to deal with that claim or any other payment issue. If funds are withheld, the Owner will have money to pay you if and when you prevail in your dispute. The General does not get paid, and so there is a financial impact whenever a Stop Notice is filed. Obviously, the greater the cumulative amount of all the Stop Notices, the more that financial impact on the General will be.

Bond the Stop Notice

If you don't bond your Stop Notice in private works situations, there is no obligation for the Owner or Lender to withhold any funds. In most cases, they will hold the funds, but there are no repercussions if they are not withheld. Therefore, if you are serving a Stop Notice, bond it.

Note: Make sure that you know who the lender is for the project. You should get this information with your contract or subcontract, but if not, go to the building department and review the permit. All the key players will be identified on the application. If a lender is identified, confirm with that lender that they are, in fact, involved in the project. This is important so that all notices will get to the appropriate party. It is not unusual for a construction lender to be changed after the permit is issued, or for the lender to be incorrectly identified. There is case law that permits a good faith mistake to excuse getting the Stop Notice to the proper party, but it is far better to avoid that issue by confirming all information you need at the beginning of the project rather than arguing about why you didn't after there is a dispute.

A bond for the Stop Notice is obtained from a surety. The bond will cost you about 1 to 1.5% of the principal amount that you are bonding, typically with a minimum cost of about $100-$150. For example, if the amount you have not been paid and that you want to serve a Stop Notice on is $30,000, the bond will cost you between $300 to $450; maybe more, maybe less, but around that cost. It is money well spent.

When you provide a bonded Stop Notice, the Owner (if the Owner is funding the project) or the construction lender must withhold the amount of the Stop Notice. If they do not, they are liable to you for any amount of your Stop Notice that is owed but not paid. If you need to sue either the Owner or the lender in such a case, and you prevail in the litigation, you get your attorney fees and costs.

This assumes, of course, that there are still sufficient funds in the construction fund to cover the amount of the bonded Stop Notice when you send it. If the Stop Notice comes in at the end of the project, which is often the case, there may not be enough money left to cover the full amount of the Stop Notice. The Owner or lender do not need to make up the difference, they only have to hold the remaining amount that is in the construction fund account. Likewise, if there are a number of Stop Notices on file and there is not enough money to pay out on all of them, then the allocation of the remaining funds is done on a pro rata basis.

The Effect of Payment Bonds on Stop Notices

In those circumstances where there is a private works payment bond for the project, and the Owner or Lender gets a Stop Notice from a subcontractor/supplier or General, the Owner or Lender is not required to withhold the amount, even if the Stop Notice is bonded. The Owner or lender must, however, give notice to the subcontractor/supplier or General of its election not to withhold on the Bonded Stop Notice and it must do so within 30 days of receiving that Stop Notice. If you want a copy of the bond, provide the Owner or Lender with a stamped self-addressed envelope, and they must provide it.

Defending Against the Stop Notice

If a subcontractor or supplier files a Stop Notice, the General can bond around it, thereby releasing those funds. The Stop Notice Release Bond surety then becomes liable for the payment, rather than the construction lender or the Owner. The amount of any Release Bond is 125% of the Stop Notice amount. The cost of the Stop Notice Release Bond is typically 1-1.5% of the amount bonded, so it can get pricey if there are a lot of Stop Notices filed or if the amount in dispute is high.

It is a good idea to contact the Owner or Lender before you post the bond, so that you are certain that the funds will be released. If there are other issues on the Project that will prevent the release of the funds, then it makes no sense to purchase a bond that will not do anything.

The Need for an Accurate Stop Notice

The Stop Notice must be reasonably accurate about the amount that is to be withheld. The Stop Notice must also refer to only one project. This is so even if there is an ongoing business relationship between the sub and the General, and they have done, or are doing, other projects and money is owed on another project. The Stop Notice can only apply to payments due on a particular project. If there are other projects where payments are not being made, then a separate Stop Notice for each project must be provided. You cannot add amounts from one project into the Stop Notice on another.

The Stop Notice can only hold funds for work that has been performed, not for work that may be performed or for costs that are anticipated in the future. For example, the Stop Notice should not include projected costs. It should not include amounts for work that has not been performed, or for delay and inefficiency costs that may be anticipated but that are speculative at the time of the Stop Notice.

If a Stop Notice is not accurate, and especially if it is grossly or willfully exaggerated, it may be declared invalid or void. When it comes time for a court to decide how much the claimant is entitled to, that determination may be zero if the Stop Notice was not reasonable in the first place. The claimant will not be entitled to any pro rata share of the funds if the court determines there was a willful misstatement of value in the Stop Notice.

The party impacted by the wrongful Stop Notice may also have a claim against the party that filed the Stop Notice for the cost of any Release Bond, and possibly other damages that might result from an improper or overstated Stop Notice. After all, the Stop Notice is stopping payments, and those payments are needed to pay other contractors and suppliers. If the Stop Notice claimant is using the Stop Notice process vindictively or punitively by grossly overstating the amount owed, there can be repercussions in the form of damages awarded to the General.

Note: One interesting and unusual aspect of Stop Notices is the fact that an Owner can demand that all Stop Notices be filed at the beginning of the project. If the Owner makes this demand, then you must comply or you will lose your right to pursue Mechanic Lien rights, not your Stop Notice rights. How do you respond to this type of request? In a word, carefully. You must provide a Stop Notice with language that says that this is a conditional or preliminary Stop Notice, that it may change, and that if it changes in amount that you will advise the Owner. This is not something that an Owner will typically do, but be aware that an Owner has the right and, if they exercise that right, you need to comply. Yes, it is a bizarre rule.

Partial Release By Payment

If, after the Stop Notice is filed, some of the money owed to you is paid, you will need to provide a Partial Release of Stop Notice for the amount that has been paid. If you reduce the amount of a Stop Notice, it must be done in writing because it is valid only if it is done in writing.

There should also be a statutory Conditional Upon Progress Payment Waiver And Release form that accompanies the Partial Stop Notice Release. This is important because if the check that you are paid with is no good, you still have lien rights and can still assert a Stop Notice for the same unpaid amount.

Recall that you can file revised and amended Stop Notices during the same period of time that you can record a Mechanic's Lien. If you file a Stop Notice at one point in the project and then perform additional work that you are again not paid for, you have the right to amend the original Stop Notice or file another Stop Notice for the additional amount you are not paid. Likewise, you can reduce any Stop Notice at any time without jeopardizing your rights. If a Stop Notice amount is changed, reduced, or increased, it must be done properly and in a timely manner, or you risk compromising that entitlement.

The Stop Notice Foreclosure Action

In order to perfect your rights on the Stop Notice, you will need to file a lawsuit. The Stop Notice claim is typically part of a more involved and comprehensive lawsuit. It can be brought as a stand-alone litigation, but that is not typical. The Stop Notice foreclosure action is intended to determine your rights to the funds that are withheld. Unlike the Mechanic Lien foreclosure action, the Stop Notice lawsuit is not brought against the property but against the construction funds that are, or should be, available.

The parties to the Stop Notice foreclosure action are; 1- the lender or Owner, depending on whom you told to withhold the funds; 2- the party that you wanted the payment from in the first place, either the General or the Owner, whoever was not paying you; or 3, if there is a Stop Notice Release Bond or if there is a payment bond for the project, then the surety that issued the Bond is a required party. You do not need to name any other parties in the Stop Notice cause of action. The action is to determine your rights to the funds being withheld because of your Stop Notice.

Time Limits to File the Stop Notice Action

A lawsuit to perfect rights under the Stop Notice can be filed as early as 10 days after the Stop Notice is filed, but no later than 90 days after the last day to file a Stop Notice. The last day to file a Stop Notice will be determined by whether the Owner's records a Notice of Completion or Notice of Cessation. If the Owner does record the Notice, then a General/Prime has 60 days to file its Stop Notice and 90 days more to file suit; 150 days from completion. If no Notice is recorded, the General/ Prime has 90 days to file the Stop Notice and 90 days to file suit.

A subcontractor's time frame is slightly less if the Owner records the Notice. In that case, the sub has 30 days to file its Stop Notice and then 90 days to file suit. If no Notice is recorded, the sub has the same amount of time as the General/Prime; 90 days to file and 90 days to initiate the lawsuit. Review the discussion in the Mechanic Lien section about how the time limits apply, depending on whether the Owner does or does not record Notice, and depending on whether you are a General/Prime or a subcontractor. The same concerns are relevant for Stop Notices. (Refer to the Appendix at the end of the book for an outline of the timeframes.)

Within 5 days after filing the lawsuit, notice must be sent to all parties involved in the action informing them that the litigation has been commenced. The Notice of Commencement of Action is ordinarily sent along with the lawsuit that is being served, but sometimes, and for tactical reasons, the lawsuit might not be immediately served on the other parties, so the Notice of Commencement is sent by itself. However you do it, be sure to serve the Notice of Commencement of Action on all parties involved in your Stop Notice action, especially on the Owner or lender. This is to assure that the lender and/or the Owner continue to hold the Stop Notice funds because they know there is a lawsuit related to those funds.

If there is a Stop Notice Release Bond in place, then the bond surety is the proper party to name in your foreclosure action. If there is a payment bond on the project, and the Owner or lender have relied on that bond and ignored your Stop Notice, then name the surety in the suit and provide notice to all of them; the Owner, the construction lender, and the bond surety. You want to be paid by one of these parties; you don't care which one.

No Stop Notice action can come to trial before the conclusion of the 90-day period to file a Stop Notice. If you file your Stop Notice early, you still need to wait until the time period expires for filing Stop Notices on the project, and that time frame is dependent on the Owner's recording (or not recording) a Notice of Completion or Notice of Cessation. If no Stop Notice action is brought within the applicable time frame, then the funds that were withheld must be released.

Be sure to monitor the recording of any Notice of Completion or Notice of Cessation. Be sure to calendar the time limits for filing your action. Do NOT miss this deadline. As we discuss later in the litigation section, you can file your action to preserve your rights, and then you can amend your lawsuit to bring in other claims. If you intend to bring the Stop Notice action, however, it must be brought within a specific time, so make sure you do.

Attorney Fees

In any litigation to recover on a bonded Stop Notice, or if there is a Stop Notice Release Bond, the prevailing party is entitled to attorney fees; but only in litigation. If you are in arbitration, there is no attorney fee recovery for the prevailing party. Of course, if the construction contract has an attorney fee provision in it, then fees and costs can be awarded in any event. Fees can also be awarded under various statutes, even in arbitration, but the fee recovery in arbitration will based on the statute, not the Stop Notice action itself.

Performance Bonds and Payment Bonds

There are two types of bonds that will be discussed here: payment bonds and performance bonds. On public works projects, performance bonds and payment bonds are mandatory and must be provided. On private works, bonds are optional, but if they are in place, the timing of any claims on them can be tricky. Not only do claims need to be filed in a particular time frame, but the time to file any lawsuits that may follow such a claim will be determined by how the bond is characterized, and whether the bond is recorded or not.

Private Works Payment Bonds

In private works, payment bonds are either statutory bonds, that is, specifically found and/or required under a particular statute, or they are common law bonds, which are provided as a result of an agreement between the parties. If you want to make a claim on a private work payment bond, and if it is a statutory payment bond, then you must provide Notice of your claim to the surety. This is accomplished in one of two ways. Notice is provided to the surety if the claimant/contractor has provided a 20-day Preliminary Notice. Alternatively, if you failed to provide a 20-Day Preliminary Notice, you must provide written notice to the surety within 15 days of the recording of a Notice of Completion or, if no Notice of Completion is recorded, then within 75 days after "completion."

As you would expect, any Notice of claim needs to include the name of the project, the Bond number, your name, the General Contractor's name, and the amount of the claim. You should also include any pertinent documents that will support your claim. This is all common sense; you want to give the surety enough information to investigate and determine the validity of your claim. When the surety receives your claim, it must respond, and must provide a determination of what it will do. It can ask you for additional information, which is what will typically occur. This request will postpone the surety's time to make a decision on the claim.

Once you have provided Notice to the surety, you must follow up with a lawsuit. Timing is important, especially in private work payment bond issues. If the bond is a statutory payment bond, and/or if the Notice of Completion is recorded, your lawsuit needs to be filed within 6 months of Notice of Completion. If there is no Notice of Completion, then the lawsuit must be filed within 6 months of the completion of the project.

Recording the Payment Bond

One interesting bump in this road is where the payment bond is recorded before the project commences. Under Civil Code sections 3235 & 3236, the Owner can require a General Contractor to record the prime contract and the payment bond. The bond and prime contract are recorded at the County Recorder's Office where the project is located. The payment bond needs to be not less than 50% of the contract value and it must guarantee that all Mechanic Liens will be paid. If the contract and bond are recorded before the project starts, the Owner is only liable for the balance of the contract amount not paid, but does not need to make up any shortfall. In other words, the Owner will not need to pay more than the contract amount if the General does not pay the subs.

If the bond is recorded, the bond surety can also require a shorter time limit to file a lawsuit than the usual 6-month time from completion. Therefore, if you know that the payment bond was recorded before the project began, get a copy and read it. The bond will include the time limits to commence any action to recover against it, and any other requirements that the surety may have for filing and perfecting a claim. If nothing else, keep the bond with your project documents so that your attorney can read it, if and when it becomes necessary.

Filing an Action on a Non-Recorded Bond

If the payment bond is not recorded, the time limit to file an action becomes a bit more complicated. The time limit to file suit depends on whether the payment bond is deemed a statutory or a common law bond. If it is common law, that is, not founded in any statute but simply agreed to by contract, then you have up to 4 years to file the action. If it is statutory (provided according to Civil Code sections 3235 & 3236) but not recorded, then you will need to file your lawsuit within 6 months of completion. In most situations this is not a problem, because, as a contractor owed money on a completed project, you will not want to wait to get paid or to file a lawsuit to get paid. Plus, your claims for Mechanic's Liens and for Stop Notice claims will need to be filed at around this same time, and so all causes of action can and should be brought at the same time and in the same lawsuit.

> **Note:** All your available remedies are cumulative, not exclusive. You do not need to wait to find out how you do in a Stop Notice action or a Mechanic Lien foreclosure before you bring an action on the Bond. Keep in mind, too, that the surety cannot demand that you satisfy your bond claim out of the distribution from the Stop Notice or Mechanic Lien recovery before you come after it (the surety) for any contribution. Your rights are coextensive, and you can make demand on the bond for immediate payment at the same time you serve a Stop Notice or file a Mechanic Lien. You don't even need to bring the Mechanic Lien or Stop Notice action at all; you can simply pursue the bond. Not a good idea to waive those other rights, but it is an option that you have.

Recovery on the Payment Bond

What you can recover through the payment bond action is pretty much everything that you claim. The payment bond will pay the reasonable value of your work, and possibly delay damages if there is a time is of the essence provision in the contract; your attorney fees if they are in the contract or if you are the prevailing party in the action, or if fees are recoverable under a statute (i.e. prompt payment statutes); interest on your claim, if the claim is for a sum certain amount; and statutory prompt payment penalties if they are applicable.

If the payment bond is a common law bond, that is, if it is part of the contract requirements, but the contract is silent on any statutory reference, then any recovery may be more restricted. In that case, the contract terms will govern, and the contract language and the language in the payment bond itself will determine what you are entitled to recover.

Summary

Payment bonds on private works projects are not that common, but they do exist. It depends on the type of project and how much the project will cost. If there is a payment bond available on the project, get a copy of it. Find out whether it is recorded, and remember to track when the project is completed. Timing is just as important for bond claims as it is for Mechanic Liens and Stop Notices. If the bond is recorded, the surety can include shorter limitations periods for filing actions, and they can also provide required procedures for filing a claim. You will need to comply. And last but not least, remember you need to serve a 20-Day Preliminary Notice.

Private Works Performance Bonds

The performance bond is different. It guarantees the completion of the project. Under some circumstances, the performance bond might actually serve the same purpose as the payment bond, because the performance would arguably include the payment of liens, and therefore subcontractors might also be paid. Primarily though, the performance bond is there for the Owner and/or lender's protection in the event the General cannot or is not performing, and the project is in jeopardy of not being completed. The General can require a subcontractor to provide a performance bond to assure the subcontractor's performance on the project.

Default: Triggering the Performance Bond

In order to trigger a performance bond, there needs to be a default. Default occurs when the Owner determines that the General is in trouble and the project is unlikely to be completed. In that situation, the Owner can terminate the contract, and can make demand on the performance bond. In a default situation, the Owner will, or should, put the surety on notice that it is considering terminating the contractor before it actually terminates. If the contractor cannot cure the default within the time provided, and the Owner does terminate, the surety that issued the bond must complete the project.

Personal Guaranties & Indemnity Agreements

Before a surety provides a performance bond for any project, private or public, it will investigate the financial solvency of both the company and the owners or principals of the company. If it issues the bond, it is because there is enough collateral available to cover any expenses that the surety may incur if the bond is needed to complete the project. The surety will typically require the contractor, General or subcontractor, to assign all rights on the project, and on other projects that may be ongoing or upcoming, to it (the surety) in the event that the bond is required to complete the project. If the contractor has enough financial wherewithal to secure that potential obligation, the surety will issue the bond.

The surety will also want a separate personal indemnity agreement whereby the Owners or principals of the company agree to personally indemnify the surety for any losses or expenses if the company's assets are not sufficient. The personal indemnity agreement gives the surety the right not only to the company's contract funds for the particular project, but to the personal assets of the Principals, as well. Just as with public works performance bonds, private works bond come with lots of strings attached that are all in the hands of the surety. In short, performance bonds are costly in every sense of the word, and there is a lot at stake.

If the performance bond is called, it means the contractor has been terminated. Termination is the consequence of a material breach. It is never a good situation. For the contractor that is terminated, it is always a very expensive situation, even more so if there is a bond.

Stop Work Statute

Stopping work on a project can be dangerous. For one thing, it gives the other party to the contract a terrific defense to any claim that you may make. Any time work is stopped by one party, there will be the contention that it is unjustified, and therefore is a material breach of the contract that entitles the other side to suspend its own performance, demand your performance, or declare the contract breached and immediately sue for damages.

However, for private contract work, the Legislature has crafted a right that allows the contractor to stop work under certain circumstances. The right to stop work is actually a lot of work in and of itself, and it involves a fairly precise series of steps that need to be completed in the proper sequence, or you may inadvertently find yourself in a breach situation, the very situation that you wanted to avoid by taking the proper steps to stop work.

Here's how it works.

If you have not been paid within 35 days after submitting your progress payment, and if there are no disputes with the Owner that would allow the progress payment to be withheld, the statute allows you to inform the Owner that if you are not paid within 10 days, you will stop work. After you have given 10 days written notice to the Owner, and all lenders and subs, you also need to post a further notice in a conspicuous place on the construction site 5 days ahead of the time that you will be stopping work.

In order to avail yourself of the stop work provision in the Code, there cannot be any dispute that the payment is due and owing. Typically the Owner will have some reason for not paying you, but assuming it is simply a cash flow issue that the Owner has and not a problem with your work, this stop work remedy is available. Plainly this factual prerequisite is going to be difficult to satisfy in most circumstances.

If you properly send the notice, and then properly post the notice on site, and if there is no dispute that you are owed the funds, you can stop work after the 10-day period. If you exercise this right, and do it properly, you have no liability for delays that the stop work may have caused. There is also a right to an expedited court hearing to determine whether you should be paid.

If the Owner pays you, or the situation is otherwise resolved, you need to provide notice that you will be returning to work. This notice is provided to the Owner, lender, and subs; all the folks that you informed in the first place that you would be stopping work on the project. The notice of your return to work and the cancellation or resolution of the dispute must be posted in a conspicuous place on the project.

Using the Stop Work Option

How valuable is the right to stop work? For smaller contracts, probably not very. If you are not being paid, and if you have a good relationship with the Owner, because it's a small contract and you are likely dealing directly with the Owner, then you will know when the Owner is having financial difficulty. If that is the case, you can probably work out some agreement to suspend the work rather than going through the effort of the Stop Work notification.

The more likely situation is on larger projects when the Owner stops making progress payments. On those types of projects, the Owner will always have some stated reason, however frivolous it may be, for not paying you. In that case, you cannot satisfy the key statutory element for stopping work; there's no dispute over the entitlement to payment.

The stop work option is not something that should be done on impulse and without consultation with legal counsel. If it is done correctly, it can be effective. After all, if you are not being paid, you are financing the project for the Owner. If the Owner goes bust and there is no payment bond, you may be getting pennies on the dollar, if anything at all, and getting it long after you performed the work. Better to call the Owner's bluff. Conversely, from the Owner's perspective, if you do not satisfy all the requirements in the statute, or if you don't do the steps correctly, you may be in material breach if you pull off the project. The ball will then be in the Owner's court and you will have an even bigger problem getting paid, if you get paid at all.

Chapter 6

Home Improvement Contract Special Requirements

HOME IMPROVEMENT IS ESSENTIALLY a remodel, and it does not matter how big or small the project is in dollar value. If you take on a residential remodel, there are certain mandatory provisions that must appear in the written agreement that you must have with the homeowner. It's OK to include other terms and conditions in the Home Improvement Contract (HIC), provided they do not contradict the required provisions. (It is also acceptable to include the HIC language in contracts for other types of projects.)

Progress Payments

First, if there will be progress payments that are keyed to specific tasks, certain required statutory language must be included in the agreement, and it must be in at least 10 point font. (Below is in 12 point font)

> **"THE DOWN PAYMENT MAY NOT EXCEED $1,000 OR 10 PERCENT OF THE CONTRACT PRICE, WHICHEVER IS LESS."**

- **At Completion of the ____________________ $_______________**
- **At Complction of the ____________________ $_______________**
- **At Completion of the ____________________ $_______________**
- **At Completion of the ____________________ $_______________**
- **Upon Notice of Completion $______________________**

"The schedule of progress payments must specifically describe each phase of work, including the type and amount of work or services scheduled to be supplied in each phase, along with the amount of each proposed progress payment. IT IS AGAINST THE LAW FOR A CONTRACTOR TO COLLECT PAYMENT FOR WORK NOT YET COMPLETED, OR FOR MATERIALS NOT YET DELIVERED. HOWEVER, A CONTRACTOR MAY REQUIRE A DOWN PAYMENT."

NOTE: You may not have a Time and Material Home Improvement Contract. They are not allowed, so do not write a contract that provides for T & M. Even if the Owner wants to do the project on a T & M basis, you may not agree and you cannot have such a contract. Time & Material is acceptable in commercial contracts, but not home improvement contracts.

Extra Work, Insurance, & Workers' Compensation Requirements

The Code requires that Home Improvement contracts provide the following three advisories regarding extra work, insurance, and workers' comp. The notice regarding your insurance must also be included with your bid. These provisions are for the protection of both the Homeowner and the contractor. They also serve as great reminders to all contractors, especially those engaged in remodels, that extras and change orders need to be in writing, that insurance is necessary and should be obtained, and that you need to have Workers' Comp if you have employees.

The home improvement contract must include the following language:

NOTE ABOUT EXTRA WORK AND CHANGE ORDERS:

Extra Work and Change Orders become part of the contract once the order is prepared in writing and signed by the parties prior to the commencement of any work covered by the new change order. The order must describe the scope of the extra work or change, the

cost to be added or subtracted from the contract, and the effect the order will have on the schedule of progress payments.

Owner may not require a contractor to perform extra or change-order work without providing written authorization prior to the commencement of any work covered by the new change order.

Extra work or a change order is not enforceable against Owner unless the change order also identifies all of the following in writing prior to the commencement of any work covered by the new change order:

(i) The scope of work encompassed by the order.

(ii) The amount to be added or subtracted from the contract.

(iii) The effect the order will make in the progress payments or the completion date.

Contractor's failure to comply with the requirements of this paragraph does not preclude the recovery of compensation for work performed based upon legal or equitable remedies designed to prevent unjust enrichment.

Commercial General Liability Insurance (CGL)

The (name on the license or 'This contractor') carries commercial General liability insurance written by (the insurance company). You may call the (insurance company) at __________ to check the contractor's insurance coverage."

Workers' Compensation Insurance: (The name on the license or 'This contractor') carries workers' compensation insurance for all employees.

Recall the earlier discussion about the consequences of failing to keep your Workers' Comp information up to date with the Contractors State License Board. Failure to have proper Workers' Comp., and failure to accurately report that information to the CSLB can result in suspension of your license. Unless there is some substantial compliance issue that excuses the non-compliance, it could be that you are contracting without a license, which means you are not entitled to be paid, and worse, you need to refund all payment received. Do not try to cut corners by playing with Workers' Compensation, or any other insurance.

As of January 2011, there is a change in the law that permits various enforcement agencies to issue Stop Work orders on a project if the contractor, or any contractor on the project, is operating without Workers' Comp. The state is serious about this; you should be too.

Required Warnings Regarding Mechanic Liens

Information and warnings regarding Mechanic Liens and its consequences must also be included in the Home Improvement Contract.

The language explains to the Owner what the Mechanic Lien is, what rights the subcontractors and materialmen have if the General does not pay them, and how it may impact the homeowner.

The required information for the Mechanic Lien warning in your contract is as follows:

MECHANICS LIEN WARNING:

Anyone who helps improve your property, but who is not paid, may record what is called a mechanics' lien on your property. A mechanics' lien is a claim, like a mortgage or home equity loan, made against your property and recorded with the county recorder.

Even if you pay your contractor in full, unpaid subcontractors, suppliers, and laborers who helped to improve your property may record mechanics' liens and sue you in court to foreclose the lien. If a court finds the lien is valid, you could be forced to pay twice or have a court officer sell your home to pay the lien. Liens can also affect your credit.

To preserve their right to record a lien, each subcontractor and material supplier must provide you with a document called a '20-day Preliminary Notice.' This notice is not a lien. The purpose of the notice is to let you know that the person who sends you the notice has the right to record a lien on your property if he or she is not paid.

BE CAREFUL. The Preliminary Notice can be sent up to 20 days after the subcontractor starts work or the supplier provides material. This can be a big problem if you pay your contractor before you have received the Preliminary Notices.

You will not get Preliminary Notices from your prime contractor or from laborers who work on your project. The law assumes that you already know they are improving your property.

PROTECT YOURSELF FROM LIENS. You can protect yourself from liens by getting a list from your contractor of all the subcontractors and material suppliers that work on your project. Find out from your contractor when these subcontractors started work and when these suppliers delivered goods or materials. Then wait 20 days, paying attention to the Preliminary Notices you receive.

PAY WITH JOINT CHECKS. One way to protect yourself is to pay with a joint check. When your contractor tells you it is time to pay for the work of a subcontractor or supplier who has provided you with a Preliminary Notice, write a joint check payable to both the contractor and the subcontractor or material supplier.

For other ways to prevent liens, visit CSLB's Web site at www.cslb.ca.gov or call CSLB at 800-321-CSLB (2752).

REMEMBER, IF YOU DO NOTHING, YOU RISK HAVING A LIEN PLACED ON YOUR HOME. This can mean that you may have to pay twice, or face the forced sale of your home to pay what you owe.

NOTE: After January 2011 there is a further requirement that before you record a Mechanic Lien you must provide written Notice to the Owner that you have are doing so. If you record the Mechanic Lien but fail to provide the Notice, the Mechanic Lien is not valid. Don't forget this warning. If your contract was entered before January 2011, but you are recording the Mechanic's Lien after that date, you also need to provide Notice to the Owner that you are going to be recording the Lien.

Required Information About Complaints to the State License Board

By Code the homeowner must be provided with information about how to file a complaint and/or obtain information about a contractor from the Contractors State License Board. If the CSLB receives a complaint about a contractor, especially on a home improvement contract, it will likely investigate. If the contractor is found to be in violation of the Rules, there can be some serious consequences. You do not want to deal with the CSLB in its enforcement capacity, but you do need to make the Owner aware of the CSLB and how to contact the CSLB if that becomes necessary.

The following is what needs to be provided in your agreement:

Information about the Contractors' State License Board (CSLB):

CONTRACTORS ARE REQUIRED BY LAW TO BE LICENSED AND REGULATED BY THE CONTRACTORS STATE LICENSE BOARD WHICH HAS JURISDICTION TO INVESTIGATE COMPLAINTS AGAINST CONTRACTORS IF A COMPLAINT REGARDING A PATENT ACT OR OMISSION IS FILED WITHIN FOUR YEARS OF THE DATE OF THE ALLEGED VIOLATION. A COMPLAINT REGARDING A LATENT ACT OR OMISSION PERTAINING TO STRUCTURAL DEFECTS MUST BE FILED WITHIN 10 YEARS OF THE DATE OF THE ALLEGED VIOLATION.

CSLB is the state consumer protection agency that licenses and regulates construction contractors. Contact CSLB for information about the licensed contractor you are considering, including information about disclosable complaints, disciplinary actions, and civil judgments that are reported to CSLB.

Use only licensed contractors. If you file a complaint against a

licensed contractor within the legal deadline (usually four years), CSLB has authority to investigate the complaint. If you use an unlicensed contractor, CSLB may not be able to help you resolve your complaint. Your only remedy may be in civil court, and you may be liable for damages arising out of any injuries to the unlicensed contractor or the unlicensed contractor's employees.

For more information:
Visit CSLB's Web site at www.cslb.ca.gov
Call CSLB at 800-321-CSLB (2752)
Write CSLB at P.O. Box 26000, Sacramento, CA 95826.

Three-Day Notice of the Right to Cancel the Contract

All Home Improvement Contract must include a Notification of the right to cancel the contract. There are two required notices with required language.

First, the Owner has a right to cancel the contract within 3 days. You need to provide this information to the Owner, and you must include a form that can be used by the Owner to cancel the agreement. If the Owner exercises the right to cancel, then any deposits must be returned in full, unless there are expenses that have been incurred in the meantime.

For example, assume the agreement is signed on Monday and calls for the contractor to be ready and mobilized on Friday. In order to perform, the contractor orders materials and puts down deposits. If the contract is then cancelled but there are certain non-refundable or re-stocking charges that were incurred, the contractor would be entitled to use the deposit to pay those costs, with the remaining balance being refunded to the Owner.

Obviously, it is advisable to hold off on any expense that is not absolutely needed until the 3-day cancellation period has expired. If there are expenses that are unavoidable, then have the Owner make those payments directly. You, as the contractor, are entitled to an advance of up to 10% of the project cost, but you should wait until the 3-day rescission period has passed before using those funds, if you can.

You must include the information that the Owner can cancel the contract in the agreement itself, and you must also provide a separate document that is the actual Notice of Cancellation. The language in the contract can be something like the following:

Three Day Right to Cancel

The law requires that the contractor give you a notice explaining your right to cancel. Initial the checkbox if the contractor has given you a 'Notice of the Three-Day Right to Cancel'. ______________ (please initial)

A Three Day Notice of Cancellation is provided with this Agreement as Exhibit A.

The following is the language for the actual and separate Three Day Notice:

EXHIBIT A

Three Day
Notice of Cancellation
(Date of transaction)

You may cancel this transaction, without any penalty or obligation, within three business days from the above date.

If you cancel, any property traded in, any payments made by you under the contract or sale, and any negotiable instrument executed by you will be returned within 10 days following receipt by the contractor of your cancellation notice, and any security interest arising out of the transaction will be canceled.

If you cancel, you must make available to the contractor at your residence, in substantially as good condition as when received, any goods delivered to you under this contract or sale, or you may, if you

wish, comply with the instructions of the contractor regarding the return shipment of the goods at the contractor's expense and risk.

If you do make the goods available to the contractor and the contractor does not pick them up within 20 days of the date of your notice of cancellation, you may retain or dispose of the goods without any further obligation. If you fail to make the goods available to the contractor, or if you agree to return the goods to the contractor and fail to do so, then you remain liable for performance of all obligations under the contract.

To cancel this transaction, mail or deliver a signed and dated copy of this cancellation notice, or any other written notice, or send a telegram

Contractor Name
Address
Not later than midnight of_______________ (date)
I hereby cancel this transaction
__________________________ (name and signature of Owner)

Seven Day Right to Cancel

The second required notice allows the Owner to cancel the contract in the event there is a catastrophic event. This is a bizarre bit of legislative meddling that is both poorly drafted and poorly considered. How likely is it that after a catastrophic event an Owner and contractor would not agree to forego the project at least for the time being? Of course, if the Owner calls and asks you to take emergency remedial action to avoid some calamity, and you do, the contract can be cancelled within 7 days and you may have an issue of payment that you need to deal with- no good turn goes unpunished after all!

In any event, and however unusual but well intentioned it is, the Home Improvement Contract must have the following language. Again, it must be in the body of the contract itself as well as included as a separate Exhibit to that agreement.

The following language should be included in the contract itself.

Seven Day Right to Cancel: The law requires that the contractor give you a notice explaining your right to cancel. Initial the check-box if the contractor has given you a 'Notice of the Seven-Day Right to Cancel'. ______________ (please initial)

A Seven Day Notice of Cancellation form is provided with this Agreement as Exhibit B.

The following, or something substantially similar, is the language for the actual and separate Seven Day Notice that should be included.

EXHIBIT B

Seven-Day Right to Cancel

[For the repair or restoration of residential premises damaged by any sudden or catastrophic event for which a state of emergency has been declared by the President of the United States or the Governor, or for which a local emergency has been declared by the executive officer or governing body of any city, county, or city and county.]

You have the right to cancel this contract within seven business days. You may cancel by e-mailing, mailing, faxing, or delivering a written notice to the contractor at the contractor's place of business by midnight of the seventh business day after you received a signed and dated copy of the contract that includes this notice. Include your name, your address, and the date you received the signed copy of the contract and this notice.

If you cancel, the contractor must return to you anything you paid within 10 days of receiving the notice of cancellation. For your part, you must make available to the contractor at your residence, in substantially as good condition as you received it, any goods delivered to you under this contract or sale. Or, you may, if you wish, comply with the contractor's instructions on how to return the goods at the contractor's expense and risk. If you do make the goods available to the contractor and the contractor does not pick

them up within 20 days of the date of your notice of cancellation, you may keep them without any further obligation. If you fail to make the goods available to the contractor, or if you agree to return the goods to the contractor and fail to do so, then you remain liable for performance of all obligations under the contract."

To cancel this transaction, mail or deliver a signed and dated copy of this cancellation notice, or any other written notice, or send a telegram to

Contractor Name
Address
Not later than midnight of_______________ (date)
I hereby cancel this transaction
___________________________(name and signature of Owner)

Last Issues for Home Improvement Contracts

Last but not least, the parties need to sign and date the agreement. If there are additional documents that are attached to the agreement, these should be identified, and there should be some language that incorporates them into the agreement by reference. There may be other provisions that you or the Owner will want to include, but the provisions identified above are required for all home improvement contracts.

If you create your own Home Improvement Agreement, you should spend a little time and money with an attorney who is familiar with contracts, especially construction contracts for home improvements. The law changes, the requirements change, and the pennies you save by not having your agreement reviewed may not be worth the dollars you spend or lose later because you did not have a proper and compliant Home Improvement Agreement.

Chapter 7

Public Works Contracting

PUBLIC WORKS CONTRACTING IS to private works contracting what rugby is to football; when you watch either game you more or less know the rules and the idea of the game, but then there are those little differences that make all the difference. You can only move the ball forward by passing backward in rugby, not exactly the same as the NFL. Neither sport is easier or less complicated than the other, and both have similar rules, but if you don't know how to play the game, you can get hurt really badly. The same can be said about Public Works and Private Works contracting; you need to know the rules or things can become really bad, really fast.

Unlike private works construction, public works has its own Code — the Public Contract Code — as well as the Government Code, both of which have specific application to public works contracting. The Public Contract Code is two volumes dedicated to only public contracting. It is complicated and it is convoluted; sometimes you need to read multiple sections together in order to determine what you can and cannot do.

When private works slows down in a recession, public sector work is usually the one area where projects are still put out to bid. Public agencies have bond money that must be spent on specific projects, and so public agencies will have funding available even in the worst of times. Public works contracting is not the same as private works, though, and those differences should not be underestimated. This section is intended to explain the rules and what you need to know about the world of public works contracting.

The Public Contract Code

The Public Contract Code (PCC) is the bible for public works projects. It is in two volumes and, after an introductory section that is applicable to all public contracts, it divides into three general sections: 1- the State of California; 2- the University of California system (UC and CSUS); and 3- public entities such as cities, towns, school districts, water districts, and sewage districts. The Code is intended to apply to all public works contracts, and is intended to be read as a whole so that what is required by one section is required by all other sections, unless the particular section specifically says otherwise.

The PCC lays out the requirements for all aspects of public contracting; bidding, awarding contracts, substitution of subcontractors, progress payment, making claims for additional compensation and/or time, for any number of things. However, certain requirements will change, depending on who the Owner is on the project. If the project is done for the State, there may be different requirements than if the project is for the City of Napa or the County of Los Angeles, and those requirements may be different still from those for a project for UCLA.

One important difference that we will discuss later is the requirement to file a statutory Government Code Claim before you file a lawsuit. If the project is for the University of California, you do not need that formal Claim. If the project is for a local agency, it is usually mandatory, unless other requirements are stated in the contract itself. As a general rule though, the PCC applies to all projects, regardless of the Owner.

The PCC is not an especially complicated Code, but it is not an easy read. If you're going to do public contracting it's worth the time to become at least generally familiar with it. If you find yourself in a situation where you need to interpret the PCC, you need to talk to an attorney. Your understanding and interpretation may be correct, but if it isn't, there can be serious consequences to being wrong.

Bidding on Public Projects

Public Owners put out bid notifications in various trade newspapers or other media so that contractors who may be interested are aware of the upcoming project. The Daily Pacific Builder is one major source of upcoming projects. The Owner's notification will usually indicate when the bid is scheduled, how much the project is estimated to cost, who the architect is, who the construction manager will be, and certain other particulars.

The notice may also indicate that there will be a walkthrough of the project site or through the facility itself. The walk-through may or may not be mandatory. If it is mandatory, you need to be there and sign in, because if you don't attend and sign in, then you cannot bid on the project. It is not uncommon to have more than one mandatory walk-through to allow the Owner to get enough participation and to accommodate the schedules of the builders.

The bid notification may also include an "engineer's estimate," the Owner's estimated cost for the project. There is nothing binding, or for that matter reliable, about an engineer's estimate other than it tends to weed out a number of contractors that may think the project is too small or too big for them. At best, the engineers estimate is an educated guess, based on historic numbers that seem reliable at the time they are put together, but when bid day comes, they become irrelevant for most purposes.

Pre-Qualification

The Owner may require pre-qualification of contractors that want to bid the project. The pre-qualification process is fairly standard. Public agencies are permitted to set up their own criteria, provided it is applied fairly to all bidders and does not allow the Owner to select a preferred bidder ahead of time. The process can still be manipulated, and it is does happen on occasion, but for the most part it is a fair process.

The form questionnaire that's used for the pre-qualification will request information about a contractor's litigation history with Owners or subcontractors, and why there was litigation. The questionnaire will also ask about Claims that have been submitted to an Owner, liquidated damages that may have been assessed, and the total value of change orders that may have been issued on the projects that are listed. The point of the prequalification process is to find out if the contractor is a problem to deal with and whether they will deliver the project on time and for the contracted price, or close to it.

The questionnaire will also ask if the contractor has ever been debarred or terminated for cause on a project. If the answer is yes, the contractor will likely be rejected. "Debarment" is when a public agency terminates a contractor's right to even bid a job for a particular agency for a specific period of time. The contractor will have done something very, very bad and the penalty is the banishing (and blacklisting) of the contractor from that agency's list of bidders. It is not a good thing at all. Even if the debarment is for another agency, the effect will likely be the same; the contractor will not be pre-qualified.

Termination for cause is no less bad, but it is not an absolute preclusion to being able to bid a project. Termination for cause is when the public agency decides that the contractor is in material breach of the contract for one reason or another and then determines, by official Board action, to terminate the contract. It is a drastic procedure for both the contractor and for the agency, and has significant ramifications for all parties. More on this later, but suffice it to say, you don't want to have to report a termination for cause on the pre-qualification.

The second level of information, equally important in the evaluation process, is whether the contractor has ever done a similar type of project, and if so, how big those projects were in dollar value, size, and/or complexity. The Owner wants to know that it can rely on the contractor, and that the contractor has the experience to handle the job.

The third level of information is a check of the contractor's business and its financials. Who is the contractor's bonding company, what is its bonding capacity, what's the contractor's present financial status? This is much the same information as any private Owner might require if they were signing up a contractor for a project for the first time and, in some cases, even if they knew the contractor beforehand.

The pre-qualification process is done by a standardized rating and point system. The contractor is awarded points based on the answers it provides, and these points are then tallied. If the points are above a certain benchmark, the contractor is pre-qualified and allowed to bid the project. If not, well, better luck next time. The process is far from perfect, and it can be gamed by the Owner to get rid of contractors that it doesn't like. For the most part, though, the system is fairly applied, even if it does tend to favor the more experienced contractors.

Contractors, too, play games with this system. If the contractor is new to the public works area, and even if it is very experienced, there can be some interesting creative writing efforts that are provided on the pre-qualification form. Careful selection of projects, careful marshalling of facts, unusual but still plausible interpretations of common words and phrases, can all be employed by contractors to present an appealing picture of that contractor's work.

With the pre-qualification process completed, the contractor is listed as a General (or an approved bidder) for the particular project. The next step is bidding the project.

Bid Bonds

In order to submit a bid on a public project, a contractor will need to provide a bid bond. Bid bonds are a guaranty by a bonding company that if the contractor's bid is deemed the low bid, but the contractor refuses to accept the contract, the public Owner will have a way to make up some of the difference between that low bid and the next low bid, up to the amount of the bond.

In order to qualify for a bid bond, the surety will review your qualifications, your company assets, its track record, and your personal assets. If the surety's criteria are met, the bond will be issued. You and/or your company will be the principal on the bond. The public Owner is the beneficiary of the bond, and the surety is, well, the surety, the obligor if you want the real technical term.

A surety issuing a bid bond may also require a personal indemnity agreement from the Owners or directors and officers of the company, especially if the company is in any way financially questionable or if the surety has no prior relationship with the company. The personal guarantee is just that- personal. You (or the company directors and officers individually) personally guarantee to pay whatever the company cannot or does not pay. The indemnity agreement usually gives the surety first dibs on your house, your assets, any profits from any other projects you may be doing; virtually anything that will secure the risk that the surety is underwriting with the bond.

Bidding

Once you have your bid bond, and after you are pre-qualified, you can submit a bid for the project. The required bid documents are included in the project manual, as well as the requirements for submitting the bid. You must comply with those requirements when you submit the bid; both in the way you submit the bid and what needs to accompany the bid.

The bid must be submitted at a specified time and in a specified place. If you miss the time to submit the bid by even a few seconds, your bid will not be accepted. No excuses, no exceptions; your bid will not be considered because legally it can't be.

The Owner accepts the bids and they are opened. The results are called out or posted by the Owner, or in some way disclosed, usually shortly after the bids are submitted. Let's assume you are lucky; the number you gave is low and you are the presumed low bid. This does not mean that you will be awarded the contract.

There are two criteria that the public agency must use to judge your bid:

1. Is the bid responsive; and

2. Are you the lowest responsible bidder.

If the public agency has pre-qualified your company, then in all likelihood you are a 'responsible' bidder; that is, you possess the necessary experience, qualifications and capability to perform the work. The more important criterion at bid time is responsiveness; was the bid that you submitted compliant with the bid requirements?

Your bid must be responsive to the bid document requirements. For example, if the bid documents require that you list a group of alternates in one part of the bid and include another group of alternates in your base bid amount, you need to do it exactly that way. If you put the alternates in another area or make up your own bid form, the bid will likely be deemed non-responsive. In short, the bid must be presented in the manner the bid documents require, and it must include the information that the bid documents specify or it will be non-responsive.

A Word About Alternates

If alternates are included, the Owner must disclose how it will consider those alternates in evaluating the bids. For example, if the project is set up to bid a number of different trade packages and the contractors are allowed to bid some of the trade packages, but don't have to bid them all, the Owner must identify the criteria it will use to evaluate the different options if the entire project will be, or may be, given out to one contractor on a kind of best overall value basis. (15) This is so that the Owner cannot manipulate the bid packages and thereby exclude certain contractors so that the favored contractor gets an advantage if they bid the project in particular way.

Typically, the Owner uses the base bid to select the lowest responsible bidder and leaves open its decision to include alternates until after that initial selection process. The contractors must include the cost of alternates along with the base bid, which allows the contractors to manipulate the system to some degree, but the bid process itself will be fair, at least in determining the lowest bid price. By Code, the determination of lowest bidder must be based only on the base bid, exclusive of any alternates, unless the alternate price is required as part of the base bid.

> **Note:** Contractors will sometimes price alternates high and the base bid a little less aggressively if they believe the alternates will

be later included. Likewise, material costs that may be broken out as alternate costs, or separate line items, can be priced higher or lower, depending on whether the contractor believes the Owner may exercise its option to include or deduct certain work. The point is that you need to strategize and evaluate the way that alternates may assist you in pricing the project. Sometimes alternates can assist to make up costs that are otherwise not included in the base bid just as sometimes a deductive change order might be anticipated so that the cost of that change is less than it might otherwise be. A public contract bid is not a simple matter of taking in bids; it requires planning and strategy to maximize your profit and your likelihood of being the successful bidder. You need to plan the bid just like you will plan the work.

Subcontractor Listing

On public bids, any subcontractor whose bid is more than ½ of 1% of the total bid amount must be identified by name on the bid form. If they are not listed, it means the General is doing the work with its own forces, and if the General doesn't have the proper license or cannot self-perform the work for one reason or another, the bid can be deemed non-responsive.

There are significant penalties if a General tries to play games with the listing law. Of course, if the work is awarded to a number of different subs, and the value of that work does not come in over the listing limit, then these subs will not need to be listed. Obviously it is important to know what the listing limit is. Your preliminary estimate will provide the rough listing limit, and as you are taking in bids and assembling your final number you will know the listing limit more precisely. Make sure you identify the subcontractors that are above the ½ of 1% limit when you submit your bid.

Bid Protests

After the bids are submitted, the various bid amounts will be disclosed. The low bidder's bid and its listed subcontractors will also be disclosed. If you are second low bid or even third, look for improprieties on your competitor's bid form and look to see whom they listed for the various scopes of work. You will know what trade packages were more than ½ of 1% of the bid because you will have listed those trades on your own bid form. If your competitor has omitted one or more, that means they must self-perform the work. The question is: can they? Do they have the proper license or the proper experience?

If your competitor cannot do the work, file a protest and see whether you can convince the Owner that the bid is non-responsive, and therefore should be thrown out. Likewise, if the form is not filled out properly, protest it as non-responsive. If the subs that are listed do not satisfy the criteria in the specifications and plans, for example they need to have a specific license but they do not, protest the bid on that basis.

Bid protests are not uncommon in public contracting. There is a strict time frame to submit these types of protests, and the bid documents will tell you what those time limits are. If your bid is close to the low bidder and you want the job, you may want to look at the bid that the low bidder submitted.

The Basis for the Bid Protest

The most important criterion for a determination of non-responsiveness is, "Does the anomaly in the bid give the apparent low bidder an unfair advantage over the other bidders?" For example, can the apparent low bidder pull the bid or keep the bid based on something that is included on the bid form? If they can, there is an unfair advantage and the bid should be thrown out.

For example, assume the bid form requires that the bid price be written out in words and also written in numerical form. If the low bidder's form has one amount in words and another amount in numbers, it is arguably non-responsive and should be rejected. The contractor has an unfair advantage because it can decide to pull or not pull its bid depending on how much was left on the table.

The Owner has the right to overlook inconsequential irregularities in the bid form, and they will exercise that discretion if they can to keep the low bid in place. The whole point of the open bidding laws is to get the public the best project for the best price. Nevertheless, the public agency cannot play favorites, and the rules need to be applied fairly.

Likewise, if the public agency throws out an otherwise responsive bid on some hyper-technicality, then that is another basis for a bid protest, but one that will likely be heard in court.

If there are serious violations of fairness in the bid process, it is possible to have the issue heard by a court. First, though, you will need to exhaust all the administrative procedures to protest the bid. However, if the project is going to be awarded and the Owner is delaying action on the protest, there are injunctions that a court can issue to prevent the award of the contract. This is rare, and it is costly.

Nevertheless, if you are a close second low bid, it may be worthwhile to look over the low bid and see if there is a protest you can file. You may just end up with the project.

Withdrawing Your Bid

What happens if you submit your bid and you are low but then you realize that there is a problem with your bid? For example, assume you provide a bid that completely omits an important trade, or more likely, you put the wrong amount for that trade into your calculation and then you submit the bid with the wrong amount included. The bid you submit is materially different from the bid you intended to submit.

What can you do? You need to pull your bid. This is an important decision and it is extremely time and fact sensitive. You have very limited time to do this, five days max, from the date of the bid opening. If you find a mistake, and cannot live with that mistake, immediately notify the Owner. Immediately!

You need to fax and send a letter by certified mail, return receipt requested, to the Owner stating what you did and why it was a mistake. Provide as much information as you can about what happened and how it happened, what you intended to submit and what you did submit, and clearly state that you are withdrawing the bid that you submitted on bid day. If your mistake was in forgetting to list certain subcontract amounts, provide the bid you submitted and provide the bid that was supposed to be submitted but wasn't, and highlight the differences.

You cannot amend your bid and add to it, even if you would still be low. You either pull the bid or you live with the bid, but you cannot adjust it after the fact. You only get one bite at the apple.

The General's Computational Error ONLY

It's also important to understand that it is only the General's bid mistake that counts. In its most simplistic terms, the mistake cannot be the result of anything other than the General's failure to add or subtract correctly. If a subcontractor calls two seconds after the bid goes in and tells you that they made a mistake and want to pull their bid, that's too bad. The sub's mistake does not justify the General's pulling its bid, even if the General included that amount in its bid and the bid is therefore too low as a result. That is not a sufficient reason, and the bid cannot be withdrawn. If the General fails to accept the contract, the bid bond can be called. As for the sub, they are bound by the bid that they submitted and that the General reasonably relied on.

> **NOTE:** It is reasonableness that needs to be shown. If the sub's bid is a quarter of what the other bids are for that trade, but it is used without calling and / or confirming that the bid is correct, the sub may escape its bid obligations, but the General won't. This is the issue of promissory estoppel / detrimental reliance we discussed earlier.

In short, if a General is going to pull its bid, they must do it within 5 days of the day that bids are opened, and the General must show that it was a mistaken calculation and not a failure to read or review the plans that caused the bid to be incorrect when submitted.

Subcontractor Substitution

The General can't withdraw its bid if the sub that was listed made a mistake, but there is a process that allows for the substitution of a listed subcontractor. The substitution process is laid out in the Public Contract Code sections 4107 and 4107.5, and it needs to be followed. If it is not, the sub and the public entity both have recourse against the General.

> **Note:** It is only for listed subcontractors that the substitution process comes into play. If the sub is not listed, there is no need to follow the procedure and you are free to change out any sub in any unlisted category any time you want. The unlisted substituted subcontractor may have other remedies it can pursue against you, but the statutory one for substitution is not one of them.

In order to change a listed sub, the General makes a request to the Owner to allow the substitution. The request provides the reasons why there needs to be a change. The Owner then sends the request to the listed sub. The listed sub has a certain period of time to dispute the substitution or agree to it. If the sub agrees, then there is no problem. If the sub does not agree, the sub sends back its opposition and the Owner sets a date for a hearing. At the hearing, the Owner provides a representative to act as the judge of the request, and both sides, sub and General, get to present their arguments and evidence to support their positions for or against the substitution.

The Public Contract Code provides specific reasons for substitution. There are only a few, but they are pretty much what you would expect:

- Failure to sign the contract
- Failure to perform

- Failure to be properly licensed
- Inadvertent listing of the wrong sub
- Failure/inability to provide a bond.

These are the only reasons for substitution, but, as you can see, they are fairly broad categories.

The two reasons that are most often asserted are the failure to sign the contract and the failure to perform. If the General provides a subcontract to the sub and the sub refuses to sign it, or the sub makes significant changes to the subcontract and insists that it will only sign it with those changes included, this can be a basis for substitution. For its part, the General must provide a subcontract that requires the sub to perform according to plans and specifications, and the General cannot include other requirements that might not have been expected when the sub's bid was submitted.

If the sub is required to provide a bond, the General must have indicated that requirement in the General's advertisement for bids. If the bond requirement is not indicated, then the failure of the sub to provide a bond is not a proper basis for substitution. If the bond requirement is properly advertised and the sub cannot, or will not, bond the project, then that can provide a proper basis for substitution.

One common dispute that arises is when the sub's bid has excluded certain things that the General wants to include in the sub's scope, or that are in the plans and specifications. This kind of issue is usually resolved through the subcontract negotiation process, and is really something that should be expected, at least to some degree. However, if the term is something that is inconsistent with the plans and specifications, then there is a problem.

For example, assume that a painting sub writes in its proposal that it is excluding any brushwork for doors and casework. Assume the Project painting spec prohibits rolling of the doors and requires that those areas must be brushed. The General will typically line out the sub's exclusion and insist that the sub's scope conform to the plans and specs. If the sub refuses to sign the subcontract unless that exclusion is allowed, or unless they are paid additional money for that work, then this could provide grounds for substitution.

The substitution hearing allows both sides to present their arguments/reasons to the Owner, and the Owner will determine the case first on the issue of what is in the Owner's best interest and then on the basis of fairness. Be prepared at the hearing to present your best case and how your best case is in the Owner's best interest.

It is also necessary that the General have another contractor in mind that is ready to perform the work, so that the Owner, at the substitution hearing, can evaluate them. And of course, if you are the General, make sure the contractor you want to bring on to the project agrees to do the work. If the substitution is allowed, any added cost is something the General will have to bear; it is not something that will result in extra from the Owner. One caveat to that point though. If the Owner demands the substitution, or the Owner demands the use of a particular sub, then there is a good argument that the additional cost is an extra that the Owner should pay for.

Substitution is an available remedy to remove problem subcontractors, but it is a remedy that has procedures that must be followed. It cannot be done unilaterally. The Owner will typically allow the General to substitute under most circumstances, so the effort should be made to engage the process. If the Owner does not allow the substitution, and the sub turns out to be a complete disaster that causes delay and other problems, the General has a defense to any Owner claim for liquidated damages, at least as those might be attributed to the sub. The General may also have a claim for delay or inefficiency, based on the Owner's refusal to allow substitution.

Penalties for Improper Substitution

If the General fails to get approval for the substitution and simply hires another contractor, the Owner can take 10% of the replacement contractor's contract value as a penalty against the General. The General pays this penalty and also pays the replacement sub so the General's real cost has been increased by that 10%. In addition, the subcontractor who has been denied the substitution hearing can obtain damages in the form of its lost profit for the project; a double whammy. The ostensible non-performing sub gets paid for not performing, the Owner gets a 10% penalty, and the General pays it all.

Submittals

In public works contracting, Owners are serious about submittals. If you are the successful General, one of the first things that you will need to provide to the Owner is either a submittal log that lays out when the submittals will be provided or, more likely, an as-planned schedule that show the submittals as an item on that schedule. This is so that everyone knows when the submittals will be provided.

The usual procedure for submittals includes architect review, return of the submittal with comments, resubmission, and then either approval or another rejection. There may be a fee attached to multiple submissions of the same submittal. Therefore, the best course is to give the architect a well-organized and complete submittal.

Subcontractors are the ones that will be providing the submittals, but the General will need to review the submittal before forwarding it to the architect. If what is being provided is a re-submittal, one that has been submitted once and been returned for corrections, then it is even more important to check it and make sure that it is complete. Even if you don't know anything about the work that is included on the submittal, the revised submittal should be compared to the original submittal that was returned, to assure that the redline corrections are actually addressed. If the corrections are not made, the General needs to know why. If the sub has a problem with the corrections, these need to be submitted in an RFI. If the subs just didn't make the changes, then the General has to make sure they do. Submittals are a pain, but they are a necessary pain, so make sure they are done correctly and completely, especially re-submittals.

> **NOTE:** The work can be started without an approved submittal, but if it is, it is done at your own risk. If the architect or Owner makes a change before the submittal is approved, and even if you installed the originally specified product, you will need to take it out and install what is finally approved. This is all at the contractor's expense too, not the Owners. It does not matter if the installed product is exactly what is called for in the plans. If there is no approved submittal, then the Owner has the right to change the requirement. If the changed specification would cost more than the original specified product, the cost difference is recoverable as a change order but the installed product will need to be taken out and the new product will need to be installed. The contractor, not the Owner, pays the cost of any repair to the surrounding work that may have been damaged because of the removal of the first unapproved product.

Certain submittals are critical to the overall construction schedule and can require a long lead-time to prepare. The as-planned schedule should identify these submittals so that everyone knows the impact of delayed approvals. Because procurement will not/should not occur without an approved submittal, these types of critical submittals can impact, if not control, the critical path. They need to be monitored for timely submission and timely approval.

Submittals are also important to uncovering problems with the plans. For example, submittals are especially important for steel fabrication and for the layout of the footings. If the architectural and the structural drawings don't line up, the misalignment may only become evident when the submittals for shop drawings are being prepared. This lack of coordination may require re-design by the architect or the engineers, and the detailer for the steel fabrication may be the first one to find the problem. These problems need to be addressed immediately if the project is to remain on schedule so the contractor is not blamed for any delay in the construction that is due to the incorrectly designed plan or specification.

> **Note:** Be especially attentive when the Owner's consultants review shop drawings and make changes. Sometimes these changes are provided in the form of "comments." They are not comments though; they are changes. For example, the so-called "comments" may add steel to particular shop drawings as redline corrections; but they are not corrections, they are revisions to the original bid set drawings. If more steel is added, that means more cost. This is a change to the plan, and you may be entitled to a change order. Remember, you bid the project according to the plans provided by the Owner. You are not responsible for the plans and specifications, and you are entitled to rely on the plans you were provided. Therefore, if you will be building the project according to the approved shop drawings, and if there is extra work/cost because of the redline "comments," you need to let the Owner know.

Submittals for things as mundane as bathroom tile are also subject to review. A specification may call for a particular tile but leave the color to be determined later. When the submittal goes in for review, the designer may select the color, which is a custom color that costs 3 times the amount of the regular color. Who pays? It depends; you may or the Owner may. If the color is not something that you, as the sub or the General, should have anticipated at the time of bid, then regardless of the Owner's reserving the right to select the color, the Owner should pay for the extra cost.

Again, get the submittals in early, and make sure you read any comments that are included in the returned document, even if the submittal is marked "Approved." Things get slipped in, and you need to be aware of them because you will want to let the Owner know of any added cost or any disputed additional requirement. If you have an approved submittal, you will be expected to provide what has been approved. If there are some comments that actually change the specification, be sure to advise the Owner if the change results in an additional cost or the need for additional time for procurement and/or installation.

> **Note:** When you are preparing any submittal, be sure that you have all of the necessary licenses for that work. As discussed before, if the submittal is a necessary part of the work, and you prepare the submittal but are unlicensed at the time, you are performing the work without a license. This means that you don't get paid for the work even if you got the license on the day after you transmitted the submittal to the architect for review. You are unlicensed, and any money you receive for any work on the Project may need to be given back.

Schedule of Values

Ordinarily, the Owner will also want the General to provide an 'as-planned' schedule very soon after the contract is awarded. The as-planned schedule shows the Owner how the General intends to proceed with the work so that the project is completed within the time limits allowed in the contract. The as-planned schedule will likely be a cost-loaded schedule, so that the Owner can anticipate its cash flow needs, as can the General.

In order to get the cost loading correct, the General will need the sub to provide a schedule of values. It is not unusual for contractors to frontload the billing so that the majority of the cost for the work is paid as it is incurred or before it is incurred, rather than after the work is complete. The Owner will review the as-planned schedule and the schedule of values, and the Owner is just as likely to "adjust" the sub's SOV to reduce or eliminate any frontloading, in fact, to backload the SOV.

The SOV is important and contractors must make sure they are not financing the project and that their bills will be paid as they are incurred, or as close as possible to when they are incurred. Keep in mind, too, that the monthly payment will be reduced by retention and that the Owner will only pay for that percentage that it believes is complete on the monthly pay app. Therefore the contractors, especially subs, should be very involved in any SOV, and should make certain that they can live with the SOV that is going to be implemented.

The schedule of values is important, and the wise contractors are the ones that provide the Owner with an SOV that will provide the needed cash flow for the project. If a sub does not provide its own SOV, the General will provide one for the sub, it will be included in the cost-loaded schedule, and if the Owner agrees to that SOV, then that's the way the sub will be paid. The idea is to avoid being cash-starved on a project, and the way to prevent that is to provide a realistic and accurate schedule of values.

Prevailing Wage

California Labor Code section 1776 and California Code of Regulations sections 16400-16403 require that all workers on any public project must be paid the prevailing wage for that trade in that area at that time. The Department of Industrial Relations website (www.dir.ca.gov) posts the prevailing wage rates for all of the trades in all of the areas of California. If you are a subcontractor and are an open shop (non-union) contractor, these requirements apply to you. If you are bidding a public project, make sure you understand the rates and how much you will need to include for the workmen you will use on a project.

The prevailing wage includes an hourly wage rate, the burden on that wage amount (social security withholds and other federal and state taxes), and the benefit package that must be included. If your workers travel from an area outside of the area where you will be doing the work, you will likely need to include a per diem subsistence amount and a travel expense, as well, for any worker.

For example, if you are a Stockton contractor who will be performing work in Oakland, then you will pay the prevailing wage for Oakland plus travel and subsistence rates. If you are an open shop contractor working on a public project and you use your own workers, the benefits, and travel and subsidence, must still be paid to the Union Trust fund, even though the particular worker will likely never get that money. Price your workers correctly or it can cost you.

If you are an open shop but there is a Project Labor Agreement ("PLA") or a Project Stabilization Agreement ("PSA") on the project, one time agreements that you must enter if you want to work on the project, you may need to take workers from the Union Hall to match your own non-union workers. Even if you don't really need any workers, you will be required to comply with the Agreement. It may be a 1:1 arrangement or some other, but you will be required to use Union labor. Your workers will be paid prevailing wage, but the Union workers may have a different rate than the DIR rate. You will need to pay that rate. It's an expense that needs to be figured into your job costs.

You also need to properly rank your workers as apprentice (first year through fourth year) or journeyman, and identify them as foreman, superintendent etc. Each of the categories has a specific prevailing wage rate associated with it. If you don't rate the worker correctly, you will be penalized. If you are an open shop and you are using labor out of the local Union Hall, be very clear about the status of the worker you are provided.

You must also be clear on what the proper apprentice to journeyman ratio is for the project, and how many of each classification you will need to perform the work. There can be penalties if the ratio of journeymen to apprentices is not met on the project. For any union worker obtained from the Union Hall, you will receive a 'release' that should have the rate and worker's status; apprentice or journeyman. Keep a copy of the release in your job file with your Certified Payroll Reports; if you are an open shop, keep all correspondence with the Union related to your requests for workers. Keep in mind, too, that if you don't like the worker sent to you, you can reject them. You will need to pay for show-up time, or for any time they work, but you don't need to keep someone just because they are sent to your job.

Note: If your non-Union crew works on a number of projects in a week, and some are public and some are private works, it may be tempting to "pay" the worker the prevailing wage for the public project but then pay them less on the private work so that, as an average, they are paid what they would ordinarily be paid for only the private work. This is a dangerous thing to do, because if the worker complains to the Union or the Owner, you can be in big trouble. A non-compliant contractor will pay the workers, the Union, and the State in the form of penalties, not to mention that the Department of Industrial Relations and the Contractors State License Board may take action. In short, this is not a smart thing to do.

Finally, be aware that if the project you are working on is being paid for in part or in total by public funds, it is subject to the prevailing wage requirements. There is case law that provides that if part of the project is funded with public money and another part is funded with other funds, the project as a whole is subject to prevailing wage requirements of Davis-Bacon and little Davis-Bacon, which means that even the part funded by private money will be subject to prevailing wage requirements.

Be very careful where the project funding comes from. If you are unclear, put in your contract with the public owner that they will pay for any required prevailing wage increases that might be assessed by the Department of Industrial Relations. This can be a very significant hidden cost. Your contract should also include some provision for wage order increases that may occur during the course of the project. If the Owner will not agree to pay these as a legitimate change order, then it is wise to include the estimated cost in your bid.

Certified Payroll Reports

The same Labor Code requirements that provide for prevailing wage also require that any contractor that provides labor at any time during the course of the project must provide Certified Payroll Reports (CPR). These are form documents that include information including the workers name, address, and social security number; the wage rate paid; any withholds and benefits that apply for each worker; the dates worked; the hours worked; the type of work performed; and the status of the worker; apprentice or journeyman.

Any contractor, General or sub, that had workers on the job for a particular week must provide Certified Payroll Reports. If the contractor did not have any labor on the job for a particular week, then a notice of exemption must be filed. If the subcontractor does not file the CPR, then it will not get paid; at least by Code it should not be paid. If CPR is not provided, the Owner can notify the DIR and the DIR can take action and penalize the non-compliant sub. If the General paid the sub, notwithstanding the lack of CPR, it can be penalized, too. The penalties are for each worker for each day that no CPR is provided, and it can add up quickly.

Forfeiture Actions

If the workers are not paid, or are paid less than what is required under the wage laws, the DIR can take action against both the sub and the General. The Owner or its labor compliance officer will notify the DIR and request a Forfeiture of wages and an assessment of penalties, assuming no CPR has been provided. If the DIR authorizes the Forfeiture action, the Owner will withhold funds sufficient to cover the penalties and wages that are due to the workers.

The DIR may also authorize "liquidated damages," which are an amount equal to the wages that are not paid. The liquidated damages are added to the workers unpaid wage as part of any recovery in the action. The Forfeiture action is usually against the General and the subcontractor. If the subcontractor is unavailable, the General is liable for the entire claim.

Responding to the Forfeiture Action

There are three things you should do if you get served with a Forfeiture action.

- Request a Settlement Hearing.
- Request a Review Hearing.
- Post a bond with the DIR for the amount of any liquidated damages.

It is strongly suggested that you do all three simultaneously, and that you do them immediately. You usually only have 60 days from the Notice of the Forfeiture action to take these actions. They are not complicated and can be accomplished quickly. If you don't do anything, the forfeiture is confirmed in the amount approved by the DIR, including any liquidated damages.

The Settlement Hearing is with the labor compliance officer for the Project or the Owner and is intended to see whether a resolution can be worked out. Usually these hearings are not productive unless the claim is undisputed, or there is no defense and you simply negotiate the payment. Nevertheless, you should use this time to demand that the Owner and/or labor compliance officer provide you with all the information they have that supports the action and that they submitted to the DIR. You should have already received this information with the DIR notice of forfeiture, but if not, and even if you have, make sure to demand it again. You want everything they have, and you are entitled to it all.

The Review Hearing Request is with the DIR hearing officer. These are a kind of mini-trial/administrative hearing, but they are not to be taken lightly. The DIR officers have significant authority, they know the law, and they try to be fair. Be prepared, be upfront, and do whatever you agree to do in the pre-hearing conferences that will precede the hearing.

> **Note:** Review the DIR website for decisions by the particular hearing officer assigned to your case. The hearing officer's decisions have no binding legal authority, but they will demonstrate the legal reasoning and thought process the hearing officer used to reach his/her conclusions, and why he/she found for or against the contractor. Read them and tell your attorney to read them. By the way, you will want an attorney in this matter, especially if the amounts at issue are significant.

Finally, and this is important, get a bond in the amount of the Liquidated Damage assessment from your bonding company. If the bond is posted, then the liquidated damages are no longer assessed. You want this bond, and it must be posted within the same 60-day time frame as the requests for hearing. If you do not post the bond and are found liable for any amount of the assessment, then an amount equal to that assessment will be added to the judgment as liquidated damages. Effectively, you will lose twice the amount of any liability if you don't post the Bond. Do not be pennywise and pound foolish! Post the bond!

Sole Source Requirements & Substitutions

A public Owner will sometimes include a specification that allows only a specific manufacturer's product for some aspect of the project. Typically these are performance-based specifications, although a particular name-brand product might also be specified. There are strict criteria for any sole source requirement, and the Owner needs to satisfy those prerequisites; the most important being that it has a need for the specific product that cannot be met by any other product.

For example, if the County has a bus wash facility that is being expanded, and that existing facility has a particular type of equipment installed, then it makes sense to continue with the same equipment manufacturer in the expanded part of the facility. However, if the County has a bus wash with one type of equipment but it is putting in a new facility in another location, then the required use of one manufacturer may not be so compelling or justified.

If there is a sole source specification, but there is a possible equal to the specified product, then the contractor must make a substitution request within a fairly short period of time after the award of the contract. The Code provides that an "or equal" substitution must be made within 35 days. If you want to offer and/or equal substitution, do it immediately!

> **Note:** Sole source specifications are not common, but where they are used they can be problematic. For one thing, these kinds of specifications provide the sole source manufacturer/supplier with something of a windfall, and one with only limited, if any, downside. Once the sole source provider has the contract, it is free to

perform as it wishes and when it wishes, and it is not uncommon for the sole source provider to "re-negotiate" the terms of its proposal after the contractor has been designated the low bidder and after the 35-day period has passed.

Owner Reputation

Certain public entities are notoriously difficult to deal with even in the best of times. It is not unusual to see projects that are nearly the same in type and scope priced dramatically differently from one city to another, even when they are adjoining cities. This is the industry's risk analysis being played out in the bids. Some public entities have reputations for being slow paying, untrustworthy, litigious, and/or all of the above. Contractors that have been in the public construction arena for a while will price their work to those agencies accordingly.

Be aware of the reputation of the Owner for the public project you are bidding. If you are a General and have not bid to this Owner before, ask your subs or other Generals that have dealt with them. You may just decide to pass on the project, or add a little to the bid to cover the added risk involved.

Progress Payments

The Public Contract Code makes it a matter of statewide concern that contractors working on public projects must be paid. The Public Contract Code says that public agencies are expected to set the standard for contracting in the state. Does that mean if you work on a public project that you will always be paid for your work? Not necessarily. Does it mean it mean that if you contract with the state that you will be treated fairly and everything will be done on the up and up? Not necessarily. Does it mean that payments will be never be used as leverage to get you to do things that are not in your contract but that need to be done anyway? Not necessarily.

Contracting with a public agency is much the same as contracting with any other owner. The provision in the Public Contract Code is important, because it is a mandate by the Legislature that contractors should be paid in a timely manner. If the Legislature thinks it's important enough to put it in the Code, the Public Owner is, at least in theory, expected to be fair and above-board in dealing with contractors. In theory!

> **Note:** The California prompt payment statutes require that all contractors, on any public and private project, must be paid promptly for their work, and if they are not, there can be serious consequences. The Codes that apply in private works situations often have the same language as the Public Contract Code. The prompt payment statutes all include, in one form or another, penalties, interest, and attorney fee provisions for late and/or unjustified withholding of payment. There are some minor variations, but for the most part, they are identical. Payment must be made promptly, and if it is not there are some serious penalties that can be applied.

The progress payment procedure is typically laid out in the Prime Contract and all contractors, Generals and subs, should review it and be familiar with it. More often than not, the subcontract will incorporate the payment procedure of the Prime Contract, and add a few things, to boot. This means that subs are, at a minimum, bound by the same procedures as the General, so it is incumbent on all contractors on the project to review the requirements for payment.

The Owner will require a payment application from the General, usually on a monthly basis. The payment application also needs to include certified payroll reports (this is mandatory on public works projects), conditional lien releases for progress payments for subs requesting payment, and final lien releases from subs that obtained payment on the previous draw. There can be any number of other required documents, and the General can demand that the sub provide still others that are not required in the Prime Contract. If you want to get paid, you need to comply and provide everything that is required; everything!

The amounts that will be paid are determined after the architect, the Inspector of Record, and/or the construction manager walk the project with the proposed billing from the General. Don't think that because you, the subcontractor, submitted a billing for a certain amount that you will be paid that amount or are entitled to that amount. It doesn't matter what you submit; it matters what is approved. Once the percentages are agreed, the progress billing is submitted with all the other required documents.

With the billing submitted, the Owner has 7 days to review it and let the General know, in writing, if there are any problems with the billing. If the Owner doesn't indicate any problems, then the billing is deemed accepted and it must be paid within 30 days. Barring any other reason that the Owner has for not paying it (i.e. disputed work issues or liquidated damages), the payment should be forthcoming.

Prompt Payment Penalties

If the Owner wrongfully withholds funds, progress payments or retention, penalties attach to the wrongfully withheld amounts. Wrongful withholding occurs when the Owner will not release the funds and has no good faith reason for not doing so. There must be a bona fide basis for the withholding, and the contractor should be told what that reason is.

For wrongful withheld funds, the Code provides that a penalty of 2% per month, or 24% per year, is allowed in lieu of interest. If the withhold is not wrongful, but is still withheld (in other words, there is a good faith dispute over the amounts), then interest at the statutory rate is allowed. Statutory interest is 10% or 7% depending on the circumstance.

These same penalties apply to the General Contractor that does not make payment to a sub within 10 days after receiving a progress payment from the Owner. If there is a good faith dispute, the General can withhold the payment, or some part of it. If not, the General must issue payment to the sub. If the General doesn't, the penalty or interest begins to accrue.

NOTE: The interest or penalty amounts do not compound. That is, you get 2% per month on the base disputed amount, but the penalty is not added into the base amount and then 2% calculated on the new amount. Likewise, the interest is not compounded by being added back into the base amount withheld and then recalculated for the 2% penalty.

Rejected Payment Applications

If the Owner rejects the payment application because the contractor failed to provide a proper or complete payment application, then the General is well advised to make the corrections required and re-submit. Arguing with the Owner about the completeness or correctness of the pay app may make you feel good, but it will delay the payment. Make the corrections quickly and resubmit, but if the rejection is not proper, put the Owner on notice that the delay is unjustified and entitles you to interest and/or penalties. If nothing else, it will document a potential claim that may assist you in final negotiations with the Owner.

Keep in mind that the initial (first) pay applications may have certain requirements that are slightly different than the intermediate pay applications that are submitted once the project is underway. The first pay applications may require schedules, shop drawings, and submittals that would not be required in later applications. The intermediate payment applications will be made on a month-to-month basis. Presumably, these become somewhat standard with only the amount changing as the project progresses.

NOTE: If the project is for the State and over $25M in contract value, you are allowed to bill for mobilization at the beginning of the project on a statutorily specified percentage basis. The mobilization costs on a large project can be significant, and by Code you are allowed to recover them during the initial billing cycles. The Owner cannot refuse to pay this cost, so bill it!

Good Faith Disputes & Penalties

In order to withhold payment, there must be a good faith bona fide dispute. Good faith and bona fide are terms of art, but are generally understood as real disputes and not pretexts. If the dispute is between the Owner and the General, then it is likely fairly clear what it is about and what the bases of the dispute are.

If the dispute is between the General and a subcontractor, the issues may not be so clear. It behooves the General to make it clear. If the General is going to withhold funds, a letter to the sub outlining the reasons for non-payment should be sent. The subcontractor should respond and present any defense. There should also be a history of the dispute documented in the General's file and the sub should be aware of the basis for any dispute, so that all reasons for the withheld payment should be clear and understood. This does not guarantee that a court will find there was a good faith basis for the withholding, but it is better than not having anything in the file and simply not paying the subcontractor.

If there is a good faith dispute between General and sub over the work the sub did, or did not, perform, or if there are back charges that are assessed, the General can hold back 150% of the cost of that disputed work from the subcontractor's payment. If there is not a good faith basis for the dispute, the General risks liability for the wrongfully withheld amount of non-payment; a 2% per month/24% per year penalty on the wrongfully withheld amount.

The Owner can withhold 150% for any disputed amount, and is likewise subject to the same penalties for any wrongfully withheld progress payment.

The fact that there is a dispute, even if it is well documented, does not necessarily mean that it will be found to be in good faith. It is possible that part of the withheld amount may have been held in good faith, but there may be a part that is not and that could result in an assessment of penalties or interest, which then triggers an attorney fees provision.

Note: If there is a dispute on some, but not all of the payment, the General might consider releasing the funds that are not disputed even though by statute it is allowed to hold 150% of the disputed amount. As an evidentiary matter later on at trial, or even at mediation, this might work to the General's advantage by further establishing the good faith/bona fide nature of the dispute. It will also reduce any penalty amount because the withheld amount will be less.

Liquidated Damages as Wrongful Withholds

It is not uncommon for liquidated damages to be assessed at the end of the project as tactic to force a reduction in the contractor claim amount. If the project is late, it does not necessarily follow that the contractors caused the delay, or, for that matter, that the project was actually delayed. If change orders are still being negotiated and time extensions are being requested and part of that negotiation, the project may actually not be late at all.

Moreover, if the Owner caused the delay to the project, or if change orders and time extensions were not properly and timely issued, then the assessment of LDs should be construed as wrongful withholdings that subject the Owner to penalties and/or interest. The Owner must have a bona fide reason to withhold even LDs, and if the assessment is not made in good faith it is a wrongful withhold.

If the delay is determined to be concurrent, caused by both the Owner and the General, and the Owner knew or should have known that it was concurrent, then the assessment of LDs and subsequent withholding is also arguably wrongful, so that penalties should attach.

Whether interest can be assessed for amounts withheld as LDs that are held in good faith, but eventually shown to be unwarranted, is an open question. In any case, the courts look at the factual bases for the dispute and will make a determination on the good faith of the dispute (or not) based on the circumstances that caused the delay, not just the fact of the delay itself.

Final Payment

Some of the things that may be required as a prerequisite to receipt of final payment are the complete Operation and Maintenance (O & M) manuals, training of the Owner's staff in the operation of the various systems, warranties and extended warranties, completed punchlist items, as-built drawings, as-built schedules, Consent of Surety releases, subcontractor final lien releases, Stop Notice releases, and release of all claims documents. As you can see, there can be a lot of effort required before the final payment needs to be made.

The final close-out documents will also need to be reviewed and accepted by the Owner and/or the architect and engineers. If the final documents don't comply, or if there are problems with the documents, they will be sent back for corrections. It is not unusual for the Owner and architect to hold the documents until everything is submitted, and then review them, sending certain ones back for correction. Until and unless all final submittals are approved, there will not be any final payment.

> Note: It is often the case that the Owner will receive the final documents, but not respond to them. You need to keep on top of this, because until those documents are reviewed and approved, you will not be paid. If a final document is not accepted, do NOT accept only a copy of the page that requires the correction. Get at least one copy of the entire plan set, or the submittal, back and review it so that you know what has been commented on and what else may need to be corrected.

Retention

Retention is that part of the contractor's progress payment that the Owner withholds until the end of the project. The Owner has a right, and in public contracts, an obligation, to withhold these funds. Retention is to assure that the project will be completed, and to make sure that there will be money available to make whatever corrections may be needed.

It can also be leverage that is used to force contractors to provide more than might be required, and as a negotiating tool to reduce end of project claims. Until it is released, the Owner controls it, and the Owner is not going to be shy about exploiting that circumstance.

One thing to be perfectly clear about is that retention is the contractor's money; it is not the Owners. In some cases, attorneys for public agencies argue to judges who are not familiar with contracting that the retention is still the Owner's money and therefore the Owner can access that money whenever it wants. Not so! Retention is the contractor's earned money and is owed to the contractor; it is just not paid at the time of the progress payment. It is not the Owner's money; be very clear on this point.

> NOTE: The percentage amount of retention the General can withhold from the subcontractor's progress billing is generally the same amount the Owner withholds. One exception: if the General advertises that it will require a bond from all subcontractors but the listed sub cannot or will not provide the required bond, by Code the General is permitted to withhold more retention than what the Owner withholds.

Securities In Lieu of Retention

The Public Contract Code provides an alternative to withholding retention from the progress payment. If the General wants to post securities in lieu of retention, it can. For example, the General can provide municipal bonds in an amount that is equal to the amount of retention for any progress payment. The bonds are placed with an escrow holder, usually a bank, where they are held until the end of the project.

If the General chooses to post securities in lieu of retention, then it receives the full amount of the progress payment with no reduction for retention. The contractor therefore has the cash, although it does not have use of its securities until the end of the project. However, the General also receives the interest on the securities that are posted as that interest is accrued. It's a pretty good deal for the General, because the

General has obtained the interest on the bonds, the contractor has use of the retention amount that is released, and the General does not need to release that retention to the subcontractor until the Owner releases the securities held in lieu. The General has full use of the subcontractor's retention for the duration of the project.

Likewise, Subcontractors can elect to post bonds in lieu of retention with the General so that they do not have the actual cash retention withheld from their progress payments. This is an arrangement between the General and sub; the Owner is not involved. The sub posts the securities in the required amount, and the General must release the retention. The problem arises when the subcontractor wants to post securities but the General does not. In that situation, the General is not likely to agree to the securities in lieu arrangement because it will need to release the retention to the sub, but the General will not be getting those funds from the Owner. Instead they will have to be paid from the General's own funds, which is not likely!

Another advantage of the securities in lieu for the General is that the Owner only needs to write a letter to the escrow holder to authorize the release of the securities. This is far easier than waiting for Board or Council approval to release cash retention, and then waiting again to have the check prepared by the public agency's accounting department. Once the Board authorizes the release, the letter goes out and the securities should be returned to the General.

Release of Retention

The release of retention, cash or securities in lieu, is governed by statute, and is required when the project is complete. By statute, a project is complete

- When the Owner takes beneficial occupancy; and
- When there is a Notice of Completion.

Beneficial occupancy is defined as the use and enjoyment of the project in the manner that was intended; basically when the Owner is using the building even if there is some minor work to be done. Within 60 days of project "completion," however "completion" is defined, Public Contract Code section 7107 requires that retention be released. Once the General has received the retention from Owner, it must release the retention to the subcontractor within 7 days. If there is a partial release only, then the subcontractors are entitled to a pro-rata share of that amount, unless the Owner specifically earmarks the retention for a particular subcontractor. In that case, the subcontractor gets whatever the Owner has released, barring any disputes the sub and General may have that would prevent full disbursement.

If either the Owner or the General fail to timely release retention, there are significant penalties that can attach, unless there is a good faith dispute to justify the withholding. If the retention is wrongfully withheld, then a 2% per month penalty attaches to the amount wrongfully withheld; the same penalty that was previously discussed for wrongful withholding of progress payments.

If there is no wrongfulness, then interest at the statutory rate applies, again without any compounding. The statutory rate in this instance is now set at 10%, a lot better than any bank is giving. Finally, if there is a lawsuit to recover the alleged wrongfully withheld funds, the prevailing party in that litigation gets its attorney fees and costs. It makes no sense to play games with retention unless there is a legitimate dispute between the parties.

> **NOTE:** Even if the subcontract has a provision that the parties will bear their own fees and costs in any dispute, the statute mandates the award of penalties, interest, and fees, and the statute will supersede any contract provision.

Conditioning Release of Retention on a Waiver of Claims

It is against public policy for an Owner to condition the release of retention on the contractor's waiver of any claim to penalties and interest for wrongful withholds. Even if the contractor did agree to waive its rights, the waiver is void as a matter of public policy. In short, the Owner cannot force you to give up your right to penalties and interest in order to obtain your retention.

Legitimate Reasons to Withhold Release of Retention

The Owner will not release retention if the project is late and liquidated damages are assessed. The Owner will eventually need to prove that the delay was caused by the contractor only in order to keep the LDs, but in the meantime the retention will not be released. If the contractor does not cause the delay, or if there is concurrent delay (the delay is caused by both the contractors and the Owner), then, depending on the facts, this could be a wrongful withhold and the Owner may be liable for penalties, interest, and/or fees.

If the Owner has actual damages, damages that are specifically excluded from the liquidated damages calculation, then it can withhold retention for those costs as well. For example, if the project is late and the Owner needed to rent a temporary replacement facility, 150% of the cost of that replacement facility can be withheld.

The Owner can withhold retention if the work is not complete. If there is punchlist work, the Owner can withhold 150% of the value of that punchlist, and it can also assign a dollar value to that work. The Owner's estimate will not be conservative so it is important to complete the punchlist quickly. For any incomplete work, the Owner can withhold 150% of the value of that work.

If there are Stop Notices filed on the project, the Owner can hold 125%, and that amount can also come from retention. Keep in mind, though, that the Owner can hold retention and can also hold funds from progress payments to cover any Stop Notice claims. Retention is to assure completion of the work; the withheld progress payments can be used to support the Stop Notices, and the Owner can exercise both options concurrently.

How Owners Can Manufacture a Right to Withhold

Owners often manipulate circumstances to encourage Stop Notices that allow retention and progress payments to be withheld so that effectively the contractors end up financing the project, or the contractor's performance bond does.

Here's how this scenario plays out.

The Owner receives change order requests, and it may even pay the change orders. What the Owner will not do is provide an extension of time for the added work. If there are a number of change orders, and if they involve significant delay (time) issues, the Owner's refusal to grant time extensions will result in one of two situations. Either the contractors will need to accelerate so they complete on time, or they will not accelerate and the Project completion will be delayed.

If the latter occurs and the project is "not completed on time," the Owner can assess liquidated damages. LDs are typically assessed on a daily basis, and they can add up fast. If the LDs are $3,000 a day, which is not unusual for a medium-size commercial project, then by the end of the first month there are $90,000 in LDs; at the end of the second month LDs are $180,000, and so on. Recall, too, that the Owner may withhold 150% of any disputed amount. Pretty soon LDs exceed the amount that the General is submitting on its monthly progress payment. The Owner then informs the General that it will not make any further progress payment because the LDs exceed the amount of the requested payment.

When the General does not receive its progress payments, it may not be able to pay the subcontractors. If the subs do not get paid, they will file Stop Notices. When Stop Notices are filed, the Owner has to withhold funds; 125% of the Stop Notice amount. The withholds for the LDs and the Stop Notice can be withheld concurrently; that is, 125% of the Stop

Notice amount and 150% of the LD amount can be withheld. On its face, the Owner has a plausible basis for non-payment of progress payments.

It isn't plausible, though. The failure to give the time extension (assuming it was requested and a proper Notice was provided) resulted in the LDs, and the LDs resulted in non-payment, which then resulted in the Stop Notices. All of this is quite wrongful, but for the time being, the contractors have a dilemma: they can continue to work without being paid or they can stop work. If they stop work, the General's performance bond may be enforced and/or the General may be terminated. It is, in short, a mess, and there are no good choices.

Not all Owners are this calculated, but some are, and it does happen. When it does, the General and subs need to meet and discuss strategy. For the Owner it is win-win, at least for the time being; either the contractors finish the project and the Owner does not have to issue any payments, or the contractors walk off and the Owner goes to the surety and demands the performance bond. At the very least, there is likely going to be litigation.

Litigation is an expensive process, especially against a public Owner, so be alert for these telltale signs and consider your options. Document the issues in writing as they arise, and make sure the Owner understands that you know what is happening. If it seems that it will be something that will go further, meet with the surety. If an Owner cannot rely on the bond to finish the project, and if the contractors decide to walk off for non-payment (a last resort that needs to be well considered), there may be a limited upside to the Owner to pursue this course.

The project may be stopped, it may cost significantly more to complete, there could be huge attorney fees and costs, and it will be an embarrassing situation if the Owner is found to be acting deceptively and unfairly. Recall that the public Owner is required by statute to set the standard for contracting in California; this is hardly the standard the Legislature had in mind. Fortunately this is not a typical situation, but it can and does happen. Be alert for the telltale signs and react accordingly.

Payment Issues in Public Works

Pay If Paid Clauses–Unenforceable

The idea of pay if paid is that a sub is paid only if the General is paid; hence pay if paid. The sub and the General both bear the risk of the Owner's default, and if the General doesn't get paid, then the sub doesn't get paid either. There is a certain logic to this approach from the General's point of view, but none from the subs.

Historically, the issue of pay if paid arose in the Mechanic's Lien context; a private works remedy that does not exist in public works. This type of payment clause created a problem because a subcontractor must file its Mechanic's Lien within a specific statutory time frame after the work is completed. The sub then had to foreclose on the Mechanic Lien in order to perfect its Mechanic Lien right within another statutory time frame. If the sub was bound by the contract provision that said it would be paid only if the General was paid, then the sub had to wait for that payment to come, if at all, and that waiting period could run well beyond the time frame to file a Mechanic's Lien.

Mechanic's Liens are constitutionally guaranteed rights, and they are not taken lightly. The California Supreme Court decided that pay if paid was unconstitutional because it interfered with the right of a mechanic to assert its constitutional lien right for payment for work that it performed. The Court held that any pay if paid clause was void as a matter of public policy. In other words, you can't have such a clause.

Thereafter, the Courts of Appeals applied the same reasoning to the public contract arena. The courts held that regardless of the fact that there are no Mechanic Lien rights in public contracting, the legal reasoning would still apply; you cannot make a subcontractor's right to payment contingent on "if" the General gets paid by the Owner. According to the courts, the sub has a right to be paid regardless.

This ruling makes sense. For example, the General might be willing to forego a final payment that might only include the sub's payment, plus a minimal markup for its own profit and overhead, and therefore would have no incentive to pursue the sub's payment from the Owner. The Owner would not pay because the General was not demanding payment, and the sub could not pursue any of its rights because the General had not been paid; plainly, an unfair situation.

Bottom line, pay if paid is out on public projects, just as it is on private works.

Pay When Paid Clauses – Possible

Pay when paid provisions are not quite so clear. In this scenario, the same concern exists, i.e. that the General and the Owner can conspire to deprive the subcontractor of its entitlement to payment. In private works, this is not altogether different from pay if paid, but it is different enough that pay when paid may still be allowed, provided the sub's Mechanic Lien rights are protected (more on this in the Private Works section). In public works, because there is no Mechanic Lien right, and because of the way the courts have addressed the issue and the way the rulings have come down, pay when paid provisions are likely still enforceable, provided certain conditions are met.

The decisions that have addressed the issue have indicated that pay when paid is acceptable if the sub is only required to wait a "reasonable time" and if that "reasonable time" is specifically defined in the subcontract. As a practical matter, this is fair. The General cannot afford to fund the project for the Owner. Therefore, to expect that the General will pay the subs if the Owner does not pay the General is unrealistic, at least for some reasonable period of time.

On public works projects, the typical subcontract provision will provide, in one way or another, that the sub will get paid within a reasonable time of submitting its billing. The definition of "reasonable time" is usually stated as the time that it takes to resolve any disputes that may arise with the Owner. How this is interpreted will depend on the circumstances. If the subcontractor is involved in the dispute with the Owner, then clearly the provision would apply and the time that it takes to resolve the dispute will be however long the sub wants to fight it out with the Owner.

The pay when paid clause also implicates the claims procedure in the Prime Contract, which is something that we will discuss in more detail later. Briefly though, if the sub is billing for change order work that the subcontractor insists it is entitled to be paid for, and the Owner has rejected that claim for whatever reason(s), the sub will need to present a claim for that amount. The sub makes the claim to the General, and the General has the responsibility for presenting that claim to the Owner on behalf of the sub. All this will take time, and will implicate the "reasonable time" provision.

If there are no disputes with the Owner and the Owner pays the progress payment, then the General must make payment to the sub within 7 days or risk the assessment of penalties and/or interest for late payment. The subcontract can provide for a longer time but there is no 'reasonable time' provision here; it is all statutory.

Finally if there is a situation where the sub has no dispute with the Owner, but other subs or the General does, then the reasonable time provision allows the General to postpone payment. If the Owner is holding all of the progress payment, then the sub is stuck. If the Owner releases some, but not all of the progress payment, then the General should pay the uninvolved sub so that no Stop Notices are filed and the dispute does not expand beyond the actual issues.

Change Orders

In every public contract, there will be a change order procedure. There are variations on the theme, but for the most part the procedures all include requirements for notice, costs, approval, and, if the change order is not approved, a procedure to assert a claim. There may be different names for the various documents that are required and how the procedure to submit those documents must be implemented, but there will always be a procedure.

There will also be procedures for Owner-requested proposals for additional work. These types of Requests for Quotations or Request for Proposals, or whatever they are called in your contact, require the contractor to price a proposed change so that the Owner can decide whether it can afford and/or really wants the proposed change.

All change order procedures are theoretically straightforward, but as a practical matter, the timing, amount of detail, and the type of detail, can make the change order process an enormous and unwieldy burden. Often the requirements are 'modified' on the job so that the information that is needed can be provided in a more simplified and efficient manner. This is not without its risks, however, because if the relationship between the parties deteriorates, the fact that a Change Order or Request for Quotation was not provided in strict compliance with the procedures can become an issue and a defense to entitlement.

Likewise, if the Owner suddenly demands compliance with the original procedure, do it, and be aware that this could signal a change in your relations with the Owner. A written confirmation to the Owner that indicates that there is now a further change to the procedure for implementing change orders, even if it is just to acknowledge a return to the original contractual procedure, will document the fact that the parties had modified that procedure, it was used and accepted for a time, and that it is now changing back to the original way. It will confirm that there was a change, and it precludes the Owner from coming back later and saying that you, the contractor, never complied with the contract change procedure. You did, they knew it, they agreed, they changed it back, and you complied again.

Subcontractors should be especially careful in the change order process. Subs are usually required to follow the change order procedure in the Prime Contract, but they may not have seen that procedure. If your subcontract incorporates the Prime Contract, then make sure you get a copy of it, especially of the change order procedure. As a practical matter, subcontractors are well advised to get a copy of the entire Prime Contract General Conditions, if for no other reason than so they know what the Owner is requiring of the General.

Step One: Notice

Regardless of who is requiring the change or why there is a change, the first thing to do in any change situation is to provide Notice. The Notice will usually come from the sub and be provided to either the General or the Owner, or both. The Notice must be timely and it should be in writing, but at the very least, the sub must tell the General or the Owner that there is a problem, that there will be extra costs, and that the change may require additional time; and all as soon as possible. The General should be especially concerned because any delay-causing event to the sub is a delay-causing event to the General and that delay event may result in extra costs and the need for a time extension.

If possible, any Notice related to changed conditions, or extra costs and/or time, should be provided early enough so that the Owner, and General, can review the situation that gives rise to the additional costs before those conditions are changed, and preferably before any additional costs are incurred.

If there is a substantial extra cost involved, the General and/or sub should get written direction to proceed before doing anything. There may be a dispute later about whether the work is extra or already in the scope, so it is critically important that the dispute and/or the directive be confirmed in writing at the outset. It is also important to let the Owner know that you are proceeding under protest with the work and will be making a claim for the costs later. The Owner can dispute the claim as extra work, but it is not possible to dispute it for lack of Notice.

Step Two: Document Costs

The second step is to document the costs as quickly as they can be ascertained. If the work is done on a time and material basis, there should be signed daily tags for the T & M work. These signed tags will likely include a disclaimer that the signature only confirms the hours and the work, but is not an agreement to pay for the work. It is important to confirm that the work is done and that the subcontractor had workers on the job. Labor rates and material costs can be determined afterwards, but the fact that the workers were there and doing the work needs to be confirmed at the time.

Step Three: Submit A Change Order Request

The sub must submit a change order request to the General, and then the General must submit the sub's costs to the Owner as a request for a change order, assuming it is not something that is strictly between the sub and the General. If the work can be priced without the daily tags, then the cost and the request for a change order should be submitted together and as early as possible. If the work needs to be done to avoid a delay, the General needs to get a directive to proceed from the Owner, or else inform the Owner that there are costs that are going to be incurred and billed, and that the contractors are proceeding with the work in order to avoid delay.

What happens to the requested change order after it is submitted depends on the Owner's response, but the change order must be submitted to the Owner so that there can be a response, otherwise the contractors will have forfeited their rights to that additional money and/or time.

Know Who Can Authorize Work

On public works projects, and as discussed earlier, it is critically important to know who can authorize extra work. Do not assume that because the architect directs you to do something, and agrees that it is extra work, that you will be paid. There are only certain persons on public works projects that have the authority to approve extra work. These people should be identified in the Prime Contract. If they are not, send a Request for Information (RFI) and find out who can direct additional work.

Likewise, for subs, the subcontract may limit the authority of the General's superintendent or foreman, and they may not have authority to direct extra work. Make sure you understand who can approve and/or authorize extra work. If there is any question, call the General or the project manager, find out, and confirm the response in writing. (16)

Note: If you do get caught in a situation where you performed work at the direction of someone who did not have the authority to direct it, there are some legal theories that can be raised to avoid the forfeiture that might occur. One way is to demonstrate that the same individual directed the work in past situations, and you were paid for it. The legal theory is ratification; the party resisting the payment now had encouraged your belief by its prior behavior. They paid you before when this person directed extra work, so why not now? For a sub there is also the theory of unjust enrichment or quantum meruit; the project got a benefit that it should pay for. Be aware, quantum meruit can only be used against the General, not the Public Owner.

Step Four: Submit A Claim

Finally, there is the claims procedure, which will be its own topic. Briefly, if the change order is denied, and assuming you want to be paid for it, then it will become a claim. If the change order is denied, then the claims procedure is triggered. If you want to be paid, you need to make a claim. Review and comply with the claim procedure. If the claims procedure requires a further notice, the sub or the General, and likely both, must notify the Owner, and they must follow the contractual claims procedure and do what it requires.

Force Account & Unilateral Change Orders

Force account has a couple of meanings. Sometimes it means that the Owner has the right to do the work itself, with its own forces, within certain dollar limits, and will not abridge the Prime Contract with the General if it does.

Other times force account is used to indicate that the Owner is directing the General, and through the General the sub, to perform work that the Owner will pay for, but the Owner requires a kind of ongoing status report on the price. It is a time and material directive with a cap, and that cap can only be expanded with prior approval from the Owner.

For example, the Owner may want to pay for certain work under force account, but will limit what it will pay for that work to $5,000. If more money is needed to complete that work, the Owner will require further notice and approval of the additional cost. Keep this limitation in mind, and get the additional authorization if you will exceed the stated amount. If the limit is exceeded without notification and further authorization, the Owner may only pay up to the authorized amount and the contractor may have to eat the rest. This authorized amount is often in the small print on the directive, so look for it and be aware of it as you perform the work.

The Owner also has a right to issue a unilateral change order. This usually refers to the Owner's right to decide on its own how much it will pay for any work performed. The Owner will then issue a unilateral change order for what it decides it will pay. The balance of whatever unpaid additional cost incurred by the contractor is then subject to the claims procedure.

Beware of Change Order Language

Change orders typically contain language that provides that by signing the change order you, the contractor, are giving up any claim for related costs, including delay, inefficiency, extended overhead, and any other cost. If all that is submitted are the hard costs, then you want to keep open your right to claim for delay, inefficiency, or other related costs, whatever those may be.

It is always best to have the most comprehensive and complete change order, but that is not always possible, especially where delay and inefficiency are involved. Those additional costs may not be known or quantifiable at the time the change order is submitted, because the full impact of the change on the work or on the schedule may not yet be fully realized. If the change order is signed with the waiver language, then it will be difficult, if not impossible, to recoup those costs later. The change order form may also have language that says that if any changes are made to the document, the document will be void and will not be paid.

One way to finesse the objectionable language is to include a letter with the change order that indicates that you are signing under protest and do not intend to give up your rights to further compensation; a reservation of rights letter. This preserves your rights, because it is, for lack of a better word, economic blackmail for an Owner to insist that you must give up your right to a claim, the full amount of which may not be known at the time, and cannot predicted under the circumstances. Of course, you always have the option of simply refusing to accept the change order with the language. If the language is not taken out and the money is clearly owed, you have a right to make a claim for that amount.

> NOTE: An interesting dilemma arises here. If the contractor includes projected costs for delay and additional time, but these do not turn out to be necessary because future events mitigate the projected impact, is the inclusion of the projected costs a False Claim? The answer is not important. The fact that the question can even be asked shows that the Owner's demand is unfair, and offers the contractor only a Hobson's choice: damned if you do and damned if you don't sign.

The statutory prohibition on waiving claims only applies to Owner-generated change orders. A General's change order form may include the language, but the sub can cross it out. If you don't cross it out, arguably you have given up the claim

Just as with any contract, read the change order before you sign it. A change order is a modification to the contract, but it is a contract in and of itself. Make sure you agree to the terms, just as you would any other contract document. If you sign it, you have agreed to those terms unless you have preserved your rights to claim for more.

Claims

In the world of public contracting there is nothing more complicated or daunting than claims and the claims procedure. It is the procedure that can get you what you are entitled to, or it is something that can cost you dearly and put you out of business. According to the courts in California, the claims procedure is not to be used as "a trap for the unwary," yet that is exactly what it often becomes. Nevertheless, for claims against the Public Owner, you will have one chance to get it right; make sure you do.

What Is a Claim

A "claim" is "a separate demand by the contractor for (A) a time extension, (B) payment of money or damages arising from work done by, or on behalf of, the contractor pursuant to the contract for a public work, and payment of which is not otherwise expressly provided for, or the claimant is not otherwise entitled to, or (C) an amount the payment of which is disputed by the local agency. This is the statutory definition, and will be the one that applies. It seems fairly simple and straightforward. It isn't.

The Two Claims Procedures

In any public works contract, there will typically be two claims procedures; the contractual claims procedure and the statutory claims procedure. Unless the contract provides that the contractual claims procedure is "exclusive," then not only will you need to satisfy all the contract claim procedure requirements, but you will also need to comply with the statutory procedures. The contract and statutory procedures often have the same requirements, but satisfying the requirement in one procedure does not mean that you have satisfied the requirement in the other procedure. You need to comply with all requirements, contractual and statutory, or you may compromise your rights.

> NOTE: The requirements for the claims procedure are typically found in the General Conditions of the Prime Contract. Review this section, outline it, identify the requirements and time lines, take note of how the procedure works, understand what triggers the procedure, know what steps are needed to comply with the procedure, know who gets the claim, know how to format the claim

and how it must be delivered, what must be included, and when it must be provided. It is important; do not ignore it.

Step One: Notice of Claim

In contracts with public agencies, there are potentially two distinct Claim Notice requirements that need to be satisfied; contractual notice and the statutory notice. If the contract claims procedure states that it is "*exclusive*" and it does not *expressly* require a Government Act Claim, there will only be one Notice required, and that Notice is whatever the contract claim procedure requires. If the contract claims procedure is not exclusive (and it must expressly state that it is exclusive, otherwise it is not), then once you have satisfied all the contract claim's procedure you will still need to comply with the statutory Notice and claims procedures. You absolutely must satisfy both.

The contractual Notice procedures can be anything that the Owner wants to include and will vary depending on the project and the Owner. Typically, the contractor submits a request for additional time and/or money, the Owner denies the request, the contractor provides Notice that it is proceeding under protest, and the contractor submits Notice of its claim to the Owner. Whatever "Notice" the contract requires must be provided. Again, it doesn't matter whether the Owner knows everything about the claim, the Notice must still be provided.

Step Two: Amount of the Claim— The Statutory Benchmarks

A local agency contract must include the statutory Public Contract Code sections for claims resolution. Sometimes the Public Contract Code section is included verbatim; sometimes it is simply referenced. For local public agencies, the section is 20104.2 and 20104.4.

The statutory claim procedure in section 20104 would apply after the contract claim procedure is completed, and only if there are unresolved claims. If the remaining claim is less than $50,000, there is one procedure for the handling of the claim by the contractor and the Owner. If the claims remaining are more than $50,000 but less than $375,000 the rules are slightly different. It is important to note that the statutory procedure applies to individual claim amounts that remain unresolved, not to aggregate claims. This is where problems can arise.

For example, the *contract* claim procedure may require that all claims that remain unresolved at the end of the project be submitted as one final claim. These claims are then negotiated and may be resolved in whole or in part. The remaining unresolved claim is therefore an aggregated claim consisting of a number of individually smaller claims.

The Public Contract Code does not address aggregate claims and so it is likely that if the contract procedure requires the claims to be aggregated at the end of the project, then that aggregated claim amount would be what was implicated by the statutory scheme. If the aggregate claim is more than $375,000, then there is no statutory procedure; it is outside of the statutory scope. By definition, a claim is "a separate demand by the contractor" therefore any unresolved aggregated contract claim is just that, a separate demand, but one that is comprised of individual and separate claims. As you can see, it gets a bit confusing here so it is best to clarify the procedure with the Owner through an RFI and also consult an attorney. Any response by the Owner, of course, needs to be in writing.

Subcontractor Claims

Recall that subcontractors cannot submit claims directly to the Owner because there is no privity of contract and sovereign immunity will apply. The General must pass-through the subcontractor's claim to the Owner. By Code, any claim must be made before final payment is issued. If the subcontractor does not submit a claim and final payment is made, then the sub's only recourse is against the General. In that situation, if the claim is really one that should have been made to the Owner, the General may have a good defense to any payment. It is therefore imperative that subcontractors get their claims in to the General and demand that those claims be passed through to the Owner.

The Statutory Procedure for Claims

If the claim amounts are individually less than $375,000, there will be an informal meet and confer with the Owner. The contractor must request this meeting and the Owner must comply with the request. While it is unlikely any dispute will be resolved at this meeting, it is an excellent opportunity to gather information. At the very least, both sides should come away with a better understanding of the issues and the reason(s) why the other side is taking the position that it is.

Assuming the matter does not resolve at the meet and confer, then depending on whom the party is with the claim, the General or a subcontractor, the next step is critical.

If the contract is with a local agency, and the claimant is the General, then a Government Code Claim must be filed. We will discuss this more in the following section, but this is a mandatory filing that is simple to accomplish and devastating if it is not. The time to file this statutory claim will be dependent on when the cause of action accrued; again, something that will be discussed in the following sections. Just understand here that if you are the General, you will need to file this critical document soon after the conclusion of the meet and confer.

If the claimant is a subcontractor, and the meet and confer is unsuccessful, the next step for the sub is to file a lawsuit. A subcontractor does not, in fact cannot, file a Government Code claim. The subcontractor's only recourse is against the General. The General is therefore the named defendant, among others, in the lawsuit, and any claims in dispute are raised against the General, not the Owner. If the subcontractor wants to assert a claim against the Owner, it must do so through the General. If no lawsuit is filed, the General has no obligation to pass anything through and the subcontractor may be left unpaid.

When the General is sued by the sub, the General will have to file a Government Code claim against the Owner, assuming it does not want to be obligated to the subcontractor for the Owner's obligations.

The Government Code Claim

A Government Code Claim is necessary and required in any situation where a claim is made against a local public agency (but not the State, UC or CSUS), except in those circumstances where the contract claims procedure expressly states that the contract procedure is *exclusive*, and that exclusive procedure does not expressly state that a Government Code Claim is required. It is critical that this Government Code Claim is filed, and that it be done timely and properly. The Government Code Claim is a fairly straightforward and relatively simple document to prepare, but it has drastic consequences if it is not filed in a timely manner.

The Government Code Claim Document

The Claim form itself is a simple document and includes the basic information related to the Claim. There may be a Claim form that the Owner wants used, but sometimes that standard form is not intended for contract-based claims. In that case, you may need to prepare your own form. If all of the required information is included in what you file, the Claim form itself should not be critical. What is important is the information.

First, identify your representative for any response from the Owner and provide their name, address, and phone number. This is important; do not forget it.

Provide some particulars and background for the project; the name of the project, the project or contract number if there is one, where it is located, etc.

Identify the issues for which the claim is made; i.e. progress payments, retention, change orders, delay and impact damages, and/or subcontractor pass-through claims.

Unless the total claim amount is less than $10,000, do not include any claim total. You can state that the amount exceeds a certain dollar value, but no specific amount should be provided.

It's important to include any *pass-through* claims, and, if possible, identify the subs with those claims by name. Before preparing the Government Code claim, contact your subs in writing and by phone, ask them if they have a claim, and tell them they need to provide you with any claim information if they have a claim. It may be possible to amend a Claim before litigation is initiated, but if a specific claim is not identified in the Government claim, it may not be brought in the lawsuit. The Claim and the subsequent lawsuit must be essentially for the same issues.

Provide the names of individuals who have knowledge of the claim and the claim issues. You should include whoever you have been dealing with as the Owner's representative, the architect, if they have been involved, and any other Owner-side individuals with knowledge of the claim issues. Your project manager, superintendent, foreman (if they are involved in change order negotiations), and any subcontractor representatives should be identified. Provide names and addresses for all, and indicate who they represent.

Do NOT include attorneys or consultants names in the claim. If you do, they become, or can become, witnesses later on if the matter goes to litigation. Think strategically as well as tactically when you are preparing your claim. Attach relevant documents that will assist the Owner in understanding the Claim.

Where to File the Claim

Where you file the statutory claim is extremely important. The Government Code, which controls this claim process, identifies four specific recipients for claims against local public agencies.

- The Clerk
- The Secretary
- The Auditor
- The Board or governing body

Your claim must get to one of these enumerated recipients, or to any other contractually specified recipient. The designated recipient must actually receive your claim or it will be deemed not submitted at all. Just because the school you build is in San Diego County does not mean that San Diego County is the proper agency to receive the claim. More likely it is the San Diego Unified School District that must be notified. If the school district is not notified, then the claim is a nullity. No other agency has any obligation to send the claim on for you if you deliver it them incorrectly!

Be sure the claim gets to the correct agency. It can be fatal to the claim if it is not received in the correct time-frame. Send your claim return receipt requested, or with some other confirmation of delivery, and do not misplace that confirmation. It will be very important later on.

The Agency's Decision on the Claim

Once the proper board, department, or clerk receives the statutory claim, it must act on it within 45 days. The agency can accept the Claim, accept in part, or deny the Claim. If there is no response at all within that time, the claim is deemed denied as a matter of law.

If the Owner denies the Government Code claim, it must deny it for one of two reasons: because the Owner believes there is no merit to the claim or because the claim is late. A claim is late if it is sent after the applicable contractual or statutory limitations period has expired.

The Late Claim Application

If the Owner denies the claim because it is late, you must file a late claim application if you still want to pursue your claim. An application to submit a late claim must be filed with the Owner. The application requires that you explain why the claim was late, why it could not have been sent earlier, and why the Owner is not prejudiced by the late submission of the claim.

If the Owner denies the application for late claim, you need to go to court and seek an order to compel the Owner to accept the claim. Depending on why the claim was submitted late, the court may or may not grant the order to allow the late claim submission. Whether the court denies the late claim application or not, you will likely file a lawsuit and proceed to litigation against the Owner, but the Owner now has a very good defense to the claim; it was late, and therefore is not valid.

> **Note:** The rulings by the court of appeals are not consistent on the effect of a judicial denial of the late claim application. Some courts hold that the denial is not determinative of the claim's merit, and so the contractor may still file a lawsuit. Other courts have concluded that if the late claim application is denied, the contractor cannot file the lawsuit because the claim is time-barred. Of course, it is better to avoid this entire issue of course by filing a timely claim.

Filing the Lawsuit

If there is no response from the Owner, the contractor has 2 years to file a lawsuit. If there is a denial, then the lawsuit must be filed within 6 months of the denial. The local public agency can always send a denial after the 45 days, which will trigger the 6-month time limit to file a lawsuit from the date of that denial.

There are a couple of things to keep in mind when you are suing a local public agency on a Claim.

First, the allegations in the Complaint that you file must be substantially similar to the allegations in your statutory Government Code Claim or the contract claim that was submitted. This means that you cannot ask for one thing in the pre-litigation Claim and a completely different thing in the Complaint. The type of claim can be broadened in the Complaint to include other causes of action, and other causes of action can be alleged, but for purposes of recovery on the Claim, what is included in the Claim and what is alleged in the Complaint need to be substantially the same.

The reason for this is simple. The purpose of the statutory (and contract) Claim procedure is to provide the Owner's decision maker, i.e.- the Board or Council, with sufficient information so that it can evaluate the issues in dispute and decide whether to settle or litigate. The Owner is at a distinct disadvantage if it believes that the only claims at issue are for X, Y, & Z but when it gets sued there are additional claims for A, B, C & D. If the Claim and the Complaint are not closely aligned, a court can throw out the additional (undisclosed) claims that were not included in the original Claim. Be sure your Claim is complete when you file it so this is not an issue.

Allege Compliance With the Government Claims Act

Another critical procedural issue that you must comply with is that your Complaint must allege that you have complied with the Government Code claim procedure and/or the contract claim procedure, and/ or that you are excused from that compliance by some legal theory or fact. If you have filed a statutory claim and it has been rejected, or if you have provided a contract claim (and it is not too voluminous) and it has been rejected, attach the Claim and the Owner's response to the Claim as exhibits to your Complaint.

If you fail to allege compliance with the Government Code Act, the Complaint will be subject to attack by the Owner and you will either be required to amend the Complaint, or worse, if you have not complied and cannot amend, you will be thrown out of court.

Where to File the Lawsuit & Possible Change of Venue

When you sue a public agency, you must do so in the county where the public agency is located. In other words, if you build San Diego High School, you must sue the San Diego USD in San Diego County Superior Court.

If the Owner cross-complains against you, or if the Owner initiates the lawsuit in the first place, then the matter can be transferred to another county's Superior Court. There is an absolute right to transfer the case in this circumstance, and even if the Prime Contract says you agree to have all matters heard in the local Superior Court, the matter can and must be transferred. (17)

You have an absolute right to change venue to another county, and you should. Why? Because if you stay in the local court, the Owner will have an advantage. The jury and the judge may, and likely will, have a predisposition for the local public Owner. After all, the jury will be paying the taxes that may fund any judgment against the Owner. A jury is not likely to want to do that, especially if finding in the contractor's favor means less money in the public coffers for things like education, fire protection, and other government services. Public Owners do not usually initiate litigation so that they can retain the local venue and the significant tactical and strategic advantage it provides.

Procedure Following the Filing of the Lawsuit

Once the lawsuit is filed against a local public agency, and assuming the Claims are less than $375,000, the court can send the parties to mediation. If the matter does not settle, the next step is judicial arbitration. More on these two processes in the Litigation section, but for now understand that there is a prescribed process in the Code for resolution of Claims less than $375,000.

If the claim is more than $375,000, the contract will likely have some resolution protocol. If not, and assuming you have properly filed your Government Code Claim and complied with all contractual claims procedures, then once you file your lawsuit you are free to pursue the litigation without going to mediation or arbitration, unless this is agreed to by all parties.

False Claims Act

A Government Code 910 claim, or a claim made pursuant to the contract claim procedure, can violate the False Claim Act if it seeks money that the contractor knows or should know that it is not entitled to receive. You don't want to tangle with this Act. If you violate the FCA, significant penalties can be assessed.

There are treble damages available to the Owner. In other words, if the claim is for $150 and it is false, the penalty is $450, or three times the claim amount. Even more problematic is the possible debarment that could occur if there is a false claim; you may not be able to work for that Owner, and likely others, for a long time. The penalties are significant and the long-term effect can be even more costly.

The False Claim Act only applies to Claims, not to litigation. When you submit the Government Code Claim or your contract claim, and if you include an amount that you know is inflated, then the FCA applies. The FCA also applies when you submit an invoice or a billing that seeks more than you are entitled to receive, but it does not apply to change order requests. Change orders do not violate the FCA until you try and get paid for them; then they can become violations. The request for a change order is not a Claim, so it cannot be false and it cannot violate the FCA.

Any submission that is intended to encourage the Owner to believe that you are entitled to be paid a specific amount, and that you know or should know is overstated, could trigger FCA. The Owner does not even need to pay the amount; it is the simple act of trying to get the money from the Owner that causes the violation.

If you overstate the amount you want in your Complaint, the FCA does not apply. The litigation privilege applies, and protects a party to a lawsuit from almost all actions that are taken in the pursuit of litigation, including overstated and false claims. There may be repercussions for frivolous lawsuits, and of course you might lose and have to pay fees and costs, but the FCA does not apply in litigation.

Pass-Through Claims

Pass-through claims are costs that a subcontractor has incurred that are the result of the Owner's directives or actions, and that the Owner has refused to pay. A subcontractor cannot sue the Owner directly, so they sue the General who in turn sues the Owner on a pass-though basis. This is so for two reasons.

First, the sub has no contract with the Owner, so there is no direct obligation between these two parties. Second, the Public Owner is protected by sovereign immunity; that is, the Public Owner, as the government, cannot be sued unless it consents to be sued. The Public Owner consents to be sued by entering a contract, but the contract it entered is with the General. Therefore, the sub has no direct recourse against the Owner except through the General.

Some things to keep in mind related to pass-through claims.

1. In order to be effective as a pass-through, the General must be liable to the subcontractor for the claim. There cannot be language in the contract or in any side agreement that would free the General from liability. If the General has no liability to the sub for the claim, the Owner has no liability to the General and the claim cannot be passed through. Therefore, do not enter any agreement that would abrogate the General's liability for any claim. If you do, you also free the Owner.

2. If the General decides not to forward the sub's claim to the Owner because, for example, the General does not believe the claim has any validity, it does so at a significant risk. If it is ultimately determined that the sub is entitled to the funds, and even if the Owner is the responsible party, if the General decided not to pass the claim on to the Owner, then the General will pay the sub for that work. (18) Therefore, if you are the General, do not make decisions about the validity of any subcontractor claim to the Owner; let the Owner do that. If the sub disagrees with the Owner's decision, then the sub needs to pursue the matter by

perfecting a claim. The General should stay out of it; just pass it on and stand back!

Problems With Pass-Through Claims

One common problem with pass-through claims is when the sub refuses to comply with the claims procedure and does not give the General the required claim documentation. Claims are presented at the end of the project. As part of the closeout on any public project, there is a required Release of Claims document. If the General does not know of the sub's claim, it may not be reserved out of the Release. In that instance, the General has freed the Owner, but the sub may still want to pursue the Claim. The General is the only viable party, and the subcontractor has failed to provide Notice of the Claim. In this situation though, the sub has provided the General with a good defense.

The General is well advised to provide some notification that a final Release of Claim will be filed and that any subcontractor issues need to be identified. Likewise, the subcontractor is advised to provide timely Notice of any claim and to follow up with the General to assure that the Claim is preserved and can be prosecuted. If the Owner has the liability, both General and sub have an interest in assuring that the Owner pays that obligation.

Another issue arises when the sub's claim appears to be overstated and exaggerated. The False Claim Act becomes a concern in this situation, because the General can be liable for the treble damage penalty if the sub's claim is false. The sub should provide Verification for the Claim that will be submitted. If the sub will not verify the Claim, but insists on submitting it, and if the General has asked for, but not received, documentation for the Claim, the General can submit it with a disclaimer; basically that it is being submitted without Verification and at the insistence of the subcontractor. In this situation, the General should forego putting its own markup on the claim amount so that there is no question that the General is not participating in the Claim.

Finally, if the sub's claim is a mixed claim, with obligations to the Owner and to the General, it should be submitted to the Owner for only that part that the Owner is obligated to pay. The issues between the General and sub should be handled between them, and without the Owner.

Likewise, if the General pursues and recovers on the sub's claim, undifferentiated or distinct to the Owner, the costs of pursuing that Claim should be reimbursed to the General. Generals should and can encourage cooperation in claims processing by including a provision that allows this reimbursement of the cost of claim presentation on a kind of percentage of recovery basis.

> **Note:** Often a sub takes the position that its contract is with the General, and therefore the sub does not care who pays the Claim; Owner or General. From the subcontractors perspective this is a good and valid argument, to a point. To counter that argument, the General needs to assure that the subcontract agreement requires this differentiation, and that the subcontractor distinguishes its claims as being Owner-directed or General-directed. If the claim is undifferentiated, but the General understands that it is a pass-through to the Owner, then it can make that Claim to the Owner. Statutory penalties or interest should not be available to the subcontractor if the Claim is undifferentiated and it takes time to sort out.

Liquidation Agreements

Liquidating agreements are agreements between a subcontractor and a General. The parties agree to some arrangement that allows the General to put forward the subcontractor's claim with an understanding that at the end of the litigation process, whether it is trial or arbitration, the subcontractor will get whatever the General can obtain from the Owner for that subcontractor's claim. In the right circumstance, these agreements are immensely practical and very effective.

The liquidating agreement removes the usual adversarial relationship between the sub and the General, so that they are able to focus their efforts on obtaining what they want from the Owner, who is the truly responsible party. Without the liquidating agreement, the subcontractor and General might not be so willing to 'trust' each other at trial or arbitration, and the Owner is often able to capitalize by playing one contractor off against the other.

With the liquidating agreement, the facts can be assembled and presented in a logical manner, unhelpful facts can be explained in a consistent and plausible way, and the Owner is forced to present its own case with its own facts, and cannot rely on the squabbling contractors to distract the fact-finder. When the circumstances allow for liquidating agreements, they can be extremely effective, not to mention reducing costs of trial or arbitration. You won't need two attorneys to present the case; you only need one.

There are certain critical requirements for a setting up a proper liquidating agreement, and if they are not present the whole effort can backfire so badly that the sub or General, or both, are left completely deprived of any right of recovery. The Owner will be scrutinizing any liquidating agreement for flaws that compromise its effect, and will assert those flaws as not just defenses to the claim but to the contractor's right to be in court in the first place. These are simple but fairly sophisticated documents that require great attention to detail.

First and most important, the sub and General cannot in any way agree that the General has no liability to the sub for the sub's claim that is being passed on to the Owner. In one way or another, the General must have some continuing obligations to the sub for payment, or the sub will lose any right to seek recovery from the Owner. If the sub absolves the General of liability for its claim against the Owner, the sub has effectively given up its claim and there will never be any recovery from anyone.

Likewise, if the General pays the sub before entering the liquidating agreement, the General will not be able to prosecute the sub's claim and recover what it has previously paid to the sub. In order for the liquidating agreement to work properly and legally, there must be an obligation from the General to the sub for the amount of the claim that is allegedly owed by the Owner. If there is not, that is, if the General is somehow freed from any obligation to pay the sub, then the Owner is also freed.

This is how that inadvertently happened in one case.

The subcontractor had the concrete place and finish work on a public project. It had to repair a certain sidewalk areas around the project. The Owner continually rejected the work and insisted that it be re-done. After the third or fourth time, the sub informed the General that it wanted to be paid for all the re-work, and that it would not perform any further re-work if it was not given a change order. A change order was submitted to the Owner, the Owner rejected it, the sub performed under protest, and then the sub submitted a claim.

In the meantime, the General paid the sub for the change order work and for certain other work with the understanding that there would be a liquidating agreement and the General would then pursue the subcontractor's extra work claim along with other claims against the Owner. Unfortunately, on the back of the check was language that specifically stated that by endorsing and cashing the check, the sub was giving up all claims against the General.

Thereafter, the General and sub entered the liquidating agreement that included the disputed sidewalk work. The Owner obtained a copy of the endorsed check and presented it as evidence. The court found that the liquidating agreement was not valid because the General had no liability to the sub because of the release language on the check. The contractors could not reinstate that obligation after the fact, and therefore, because there was no liability for the General, the Owner had no liability for the claim. There was more to this case, but for purposes of this discussion, the point is that the liability of the General to the sub has to be continuing. If there is no continuing liability, the Owner has no liability. (19)

> **Note:** The General may make payment to the sub before any lawsuit against the Owner and this will not violate the liquidating agreement, provided there is no language in it that would settle and release of the sub's claims. If your accounting department puts release language on the back of checks to subcontractors, make sure it does not go on any checks that would be related to subcontractor claims.

The liquidation agreement is complicated and highly technical, and an attorney should be used to draft the agreement. There are some form agreements available, but they may need to be modified. Again, an attorney should be used to do this. If you are going to use a liquidating agreement, make sure it is done correctly.

The Liquidating Agreement

Any liquidation agreement should include a provision that identifies who will pay for the fees and costs at trial against the Owner, who will pay for the sub's expert's reports and testimony at trial, that may be needed to prove the subcontractor's claims, and what cooperation the sub will provide when the General's attorney presents the sub's claims at the trial. The agreement should include how the recovery on the sub's claims will be divided up between the sub and the General, and whether there is a guaranteed minimum that the General will provide if the recovery against the Owner is not sufficient.

Some other important considerations that are sometimes overlooked include who is responsible if the Owner proves offsets in the form of actual and/or liquidated damages that are attributable to the sub? Who pays them? Are they offset against some agreed amount? When are they taken, before or after the sub is paid, and how will any pro-rata share be determined?

There should also be a "fail safe" provision that provides that if the liquidating agreement is somehow defective and it cannot be amended to achieve its intended purpose, the agreement will be void and the parties will retain all their rights to assert their own claims individually and directly in the litigation/arbitration. As you can see, these agreements include a lot of 'what-ifs,' and an attorney who is familiar with the law, and with pass-through claims in particular, should prepare them.

Finally, if the sub has filed a Stop Notice and has initiated the foreclosure action on it, it will be important to address who will present this part of the sub's claim. The sub might want to assign the Stop Notice action to the General as part of the liquidating agreement, or it might want to prosecute it on its own. The sub might dismiss the cause of action, although this is risky if there is any chance the liquidating agreement can be undone. In any case, Stop Notices are a loose end and can end up complicating matters. How they will be handled should be addressed and decided.

The advantages of Liquidation Agreements are the cost savings obtained by presenting a united front, and the potential for a greater and more likely recovery against the Owner. These agreements are only for claims against the Owner. If the sub and the General have claims against each other, they need to decide how and when they will resolve them. Sometimes these contractor-only claims will prevent the liquidating agreement. When there is a possibility of cooperation and the facts are right, liquidating agreements are worth looking into.

Joint Defense Agreement

If the General and the sub decide to enter a liquidation agreement or some other agreement that involves the sharing of confidential information, the attorneys for the sub and the General should have a joint defense agreement. If the attorney doesn't bring this up, you should.

The joint defense agreement, otherwise known as a common interest agreement, is important for a number of reasons, not the least of which is that you do not want the Owner or any other party to be able to find out what the attorneys have been discussing with you. The issue here is the attorney-client privilege, and you want to keep that privilege in place.

If there is a liquidation agreement, the parties to the agreement will necessarily need to discuss the claims, the issues, and the facts giving rise to the claims. The attorneys for the General who will be taking the sub's claim to trial need to discuss the claim with the sub. If there is no joint defense agreement, it is possible that the parties to the liquidating agreement will be required to reveal confidential discussions. This is not a good situation.

The attorney-client privilege is at stake here, and it is an extremely important privilege. The privilege protects all communications between an attorney and the client. If you are the subcontractor, the General Contractor's attorney is not yours, and vice versa- if you are the General, the sub's attorney is not yours. Discussions with unnecessary third parties present are not protected unless that third party had to be there for some specific and case-related reason. If there is no joint defense agreement, discussions between sub and General with attorneys present is arguably not protected from disclosure.

The way to prevent this waiver of the attorney-client privilege is a Joint Defense Agreement that provides that any discussions between the parties to the liquidating agreement are privileged, are intended to be privileged, and are only taking place because the parties have entered into, or will enter into, a liquidation agreement, and therefore have a common interest and want to protect all their conversations and discussions from any disclosure to third parties.

If there are documents that will be shared and reviewed, the parties should include a provision in the joint defense agreement that those documents are only shared for purposes of the liquidation agreement-joint defense agreement and will not, and do not, lose their privileged and/or proprietary status because they have been shared under those circumstances.

Sometimes a joint defense agreement will provide that the shared documents will never be used against the other side in any subsequent litigation. This is a dangerous provision to include. There may be circumstances where those documents will be needed and would have been available through discovery. You do not want to waive the right to obtain them if the relationship deteriorates later.

Your attorney needs to consider what will be shared and what will not be shared, and will need to prepare the limiting language for the disclosure. Be sure that you are candid, very candid, with your attorney, so that your attorney knows what is available and what will actually be provided to the other side.

If the documents or the discussions between the cooperating parties are not related to the liquidating agreement or the litigation, they will probably not be privileged from subsequent disclosure. If during the discussions related to that project, you make comments or statements related to another project and another claim, those statements are not privileged, and you may have waived any privilege by talking about them with an unnecessary third party present. In short, stay focused and don't be spouting off about things unrelated to the issue at hand, especially when the spouting off contains information that is privileged.

Stop Notices and Bond Issues on Public Works Projects

Preliminary Notice

A Preliminary 20-Day Notice is notification to the Owner and General, and any others that may be required to be notified, that you are on the job, performing work and/or contributing materials, and that you expect to get paid. The Preliminary Notice is used on all construction projects, public and private, and serves the same purpose on either; notice that you are there.

> NOTE: For public works projects, if you have a direct contract with the Owner or with the General Contractor, you do not need to send a 20-Day Notice. If you are a supplier to a subcontractor or a 2d tier subcontractor, you must file the Notice to preserve your rights. There is no downside to sending the 20-Day Notice, even if you don't need to, so if you are in the habit of sending them when you get a new project, keep with that habit. It's a good one to have.

The 20-Day Notice allows a subcontractor or supplier to recover payment for work or materials that are provided to the project within the preceding 20 days and any time thereafter. For example, if you work 40 days on a project and then file your 20-day Preliminary Lien, the only right you have preserved is for the work done in the immediately preceding 20 days. This does not mean that you have no rights to claim for the work that went on before the 20-Day Notice; it just means that you cannot claim any statutory right to payment except for the work completed in the immediately preceding 20 days.

The 20-Day Notice is a very easy form to fill out, but you need to fill it out correctly. Make sure that you identify the correct owner and the correct project. There will also be a place to include the estimated value of your services. This must be a reasonable estimate; do not understate the amount or overstate it. If it is completely inaccurate, a court may deny the validity of the Notice and possibly the validity of any Stop Notice or bond claims that you want to file. The Owner and General rely on the information, and if they are misled by the "estimated value," you may be the one to suffer the consequences, not them.

Stop Notice

The Stop Notice is a notice to the Owner that the General is not paying the sub, and the sub wants the Owner to hold money from the General in a specified amount. The Stop Notice is mailed, return receipt requested, to the Owner, and to the General as well. If it is not sent to the General, the Owner will usually inform the General that it has received the Stop Notice, what the amount of the Stop Notice is, and that the Owner is going to withhold 125% of the Stop Notice amount from the General's contract amount. The General will not get paid that amount until the Stop Notice is released.

The Stop Notice is something of a two-edged sword. If the Owner holds the funds from the General, then the General will not have funds to pay the sub. If a Stop Notice is filed, it pretty much guarantees no payment to the sub. If there is a good faith dispute over the obligation to pay the sub, then the General has a defense to the payment, in any case.

Stop Notice Release Bond

If the General wants the funds released, it can bond around the Stop Notice. The General will obtain a Stop Notice Release Bond from a surety and provide it to the Owner. The bond will cost about 1.5% of the penal sum (the amount of the Stop Notice held by the Owner), so if the Stop Notice value is very high, the bond cost will be fairly high. FYI, the surety that provided the Payment or Performance Bond cannot issue the Stop Notice Release Bond. Ask your bond agent for the information on this. The Release Bond will allow the Owner to release the withheld Stop Notice funds to the General, but the General has no obligation to pay the sub that filed the Stop Notice.

> **Note:** An important caution here; the Owner does not have to release the funds, even if the General posts the Release Bond. Therefore, before you pay the bond premium, confirm with the Owner that if the Release Bond is obtained then the Stop Notice funds will actually be released. Get the confirmation in writing, too.

Partial Release of Stop Notices

A Stop Notice can be reduced after the Stop Notice is filed. If the General and the sub agree that the General will pay some, but not all, of the amount demanded by the sub, then the sub will need to send a revised or reduced Stop Notice to the Owner. The General will want the original of any Reduced Stop Notice and the sub will want a cashier's check in return. This way, the General has control of the Stop Notice Release and can make sure the Release is sent to the Owner, who presumably will be expecting it, and the sub will be sure it is paid and will not have to wait for the check to clear or bounce.

Contesting the Validity of the Stop Notice– Summary Proceedings

A Stop Notice can be, and usually is, filed for any money the claimant believes it is owed. A subcontractor or supplier can file a Stop Notice for retention that is not yet due, change orders that are not yet approved, and/or delay and impact damages that may not yet be agreed to at the time. It may be that the Stop Notice is speculative about the amount owed, or premature because certain funds are not yet available, but that does not mean it is not valid.

However, there are circumstances where a Stop Notice is not legitimate and the General can contest it.

The four reasons that a Stop Notice can be challenged are:

- The Stop Notice is filed by someone that has no Stop Notice rights (a supplier to a supplier to a subcontractor);
- It is not timely;
- It is grossly overstated;
- It is not the type of claim that is permitted in a Stop Notice.

In these circumstances the Stop Notice can be challenged in court by summary proceeding. In order to initiate this type of proceeding, the General sends a letter to the Owner and lays out the reasons why the Stop Notice should not be honored; one of the four reasons outlined above. The letter demands the release of the withheld funds. An affidavit signed under penalty of perjury by the General must also accompany this demand.

The Owner forwards a copy of the General's letter contesting the Stop Notice to the subcontractor. The sub has twenty days to respond, and if no response is provided the Owner can release the funds to the General. This is unlikely, and so the Owner will continue to withhold the Stop Notice funds.

If the General wants to pursue it further, the next step is court. The General files an action in the Superior Court where the project is located and requests an immediate hearing. These summary proceedings take precedence over other matters on the court's calendar, and so the court will set it for hearing fairly quickly. The General and the sub will then submit their legal briefs, a hearing date is set, the attorneys make their arguments, and the court rules on the case. If the judge finds for the General, the Stop Notice is extinguished and the funds are ordered released. If the court finds for the subcontractor, the Stop Notice remains in place.

As you can imagine, this process is costly and the facts have to be really outrageous for a court to extinguish the Stop Notice. The Release Bond is the far more likely and logical way to proceed. However, Summary proceedings are an option in certain cases.

Statute of Limitations to File a Stop Notice

There are important limitation periods related to Stop Notices. They are confusing, they apply only to subcontractors, and they change depending on what the Owner has or has not done at the end of the project. The following discussion of the statute of limitations is directed at subcontractors. Generals need to understand these limitations too, because the General's defenses to a sub's Stop Notice action will be impacted by a subcontractor's compliance with the limitation period. (See the matrix for this timing in the Appendix at the end of the book.)

If the Owner has recorded a Notice of Completion or a Notice of Cessation, then the subcontractor must file its Stop Notice within 30 days of the recording of either Notice.

If the Owner does not record the Notice but the project is complete, as that definition is used in public work (i.e.- beneficial occupancy or the owner accepts the project by recording a notice of acceptance), the subcontractor has 90 days from that "acceptance".

If there is no Notice of Completion and work has stopped for a continuous 60-day period, the subcontractor will have 90 days from the end of that 60-day period of cessation to file its Stop Notice if the project is done for a local public agency.

If the project is a State project, the 90-day period does not apply, and a subcontractor must still file its Stop Notice within 30 days of the 60-day cessation of work that defines completion. (This will change as of July 2012 when the State and local agencies will use the same timeline that local agencies now use.)

It is also permissible to file and release a Stop Notice, and then re-file another for the same or a different amount. This is only possible however, if the time to file a Stop Notice has not already run. In other words, if you file a Stop Notice on the project while it is ongoing, but then release it for some reason, as long as you re-file the next Stop Notice within the statutory timeframes discussed above, then the Stop Notice is valid. If you decide to release your Stop Notice but re-file it after the statute of limitations has run out, you will be out of luck.

If you want to be sure that you know when the Owner records a Notice of Completion, you can include a check for $2 to the Owner when you serve the Stop Notice. If you do this, the Public Owner is required under Civil Code section 3185 to notify you when the last day to file a Stop Notice is, based on the agency's recording of a Notice of Completion or a Notice of Cessation.

Sometimes the form Stop Notice that you purchase at the local stationary store includes language to this effect. If the language is not included, add it and enclose a check for $2 when you serve your Stop Notice. You want to know when the Owner records the Notice of Completion; it is a critical date. If you paid the $2 but the Owner fails to provide you with the information, the Owner cannot object that the Stop Notice is not filed within 30 days.

Perfecting the Stop Notice

You perfect your Stop Notice right by filing a lawsuit. The earliest you can file your Stop Notice lawsuit is 10 days after you serve the Stop Notice. The latest date is 90 days after the last date to file a Stop Notice. The last date depends on whether or not there is a Notice recorded by the Owner and if not, when the work is "accepted" by the Owner. It also depends on whether it is a local public agency project or a State project. Whenever the last day to file a Stop Notice occurs, you have to file a lawsuit within 90 days. If you miss that date, you lose the right to bring the Stop Notice lawsuit.

Within 10 days of the date the lawsuit is filed, you want to file and provide a Notice of Commencement of Action to the Owner. Typically, the Notice of Commencement is included with the Complaint that will be served on the Owner, because the Owner will be a party to the Stop Notice lawsuit. More on this later, but for now remember you want to provide this Notice because it tells the Owner you have filed an action and the Owner should not release any funds.

The Stop Notice Lawsuit

The parties to the Stop Notice action are three; the General, the Owner, and the Stop Notice claimant, usually a subcontractor or supplier. Typically the Owner has no liability in the action and is merely a stakeholder. The litigation is between the sub and the General and will determine the right to the funds and the amount that must be paid. The Owner will disburse the funds as the court determines at the conclusion of the lawsuit.

If the General has bonded the Stop Notice, the situation is a little different. The Owner is not a party in the litigation; the surety that provided the bond is, but unlike the Owner, the surety is liable for payment if the subcontractor/supplier prevails in the litigation.

There are a couple of ways the Owner can be a party even if a bond was provided.

If the Owner refused to release the withheld funds after a release bond was posted, the Owner will likely be a party in the litigation, probably named by the General but not the subcontractor.

The Owner can become a party with liability exposure if it contends that it is entitled to keep the Stop Notice funds because there are liquidated damages. If the Owner takes this position and refuses to give up the withheld funds, the subcontractor can force the Owner to prove the delay issues that give rise to the assessment of liquidated damages.

The sub may not have any direct contractual right against the Owner for any other claims, but if the Owner contests the sub's rights to the Stop Notice funds because of delay issues, the Owner must prove that delay. (20) Rare, but it does happen.

Performance Bonds

In public works contracting there are two required bonds that must be provided for any project, a performance bond and a payment bond. The General can require similar bonds from some or all of its subcontractors, but these are not mandatory.

The General's performance bond is for the protection of the Owner and assures the Owner that there will be money to complete the project in the event the General cannot or does not. Performance bonds are required by statute on all public contracts over $50,000.

An accepted and approved surety must underwrite the performance bond, and it must be for the same amount as the Prime Contract. If that contract value increases, the bond amount will typically cover that too, within limits. Any significant changes in the potential liability of the surety need to be reported to the surety, or that change may not be covered.

Owner's Demand on the Bond

If there is a problem on the project, and the Owner terminates the General, the Owner will make a demand on the surety to complete the project. This is a very bad situation and involves some very serious legal issues. The following is a brief discussion of the issues that might follow.

If the surety does not get notice of the dispute from the Owner before the claim is made on the bond, it can refuse to pay the 'penal sum' of the bond (the stated amount of the bond). Notice is a prerequisite, and failure of notice is a defense to any demand on the bond.

The surety may decide the Owner's claim is unjustified, that there was no right to terminate the contractor, and therefore there is no requirement for the surety to perform. Similarly, the General may contend that the termination is without any legitimate basis and can insist that the Surety not perform on the Bond. The Surety may agree, and it can refuse to perform.

This is risky because effectively the Surety is making the determination that the Owner is in material breach. If that determination is wrong, the Surety is in material breach. The Bond Principal, usually the General, and any personal indemnitors for the Bond are still liable for any loss, but the Surety may have its own independent liability. In short, the whole thing gets very messy if the Surety does not perform and even if it does.

Note: The Surety will require an indemnity agreement from someone other than the Principal so that in the event the Principal cannot meet the obligation to reimburse the Surety, if the bond is used to complete the project, the personal indemnitors will. There will always be some form of separate collateral indemnity agreement, either personal or some other form, that will cover the Surety's potential loss.

The Surety can ignore the Principal's request and pay the penal sum, or take over the project and perform the remainder of the contract, and then seek to recover from the Principal whatever was paid to complete the project. This, too, is risky for the Surety. If it turns out that the Owner's termination was not valid, then the Principal's obligations to indemnify the Surety may be exonerated so that the Surety is not repaid at all.

The same issues arise, and the same consequences result, for subcontractors that are required by the General to have performance bonds on a project. If the bond gets called because the subcontractor fails or goes bankrupt, the Principal on the bond is obligated to re-pay the Surety for any costs. Even if the General, as a condition of the subcontract agreement, pays for the bond, the Principal on the bond is still the subcontractor and any indemnitors that may be involved.

Payment Bonds

Payment bonds are required on any public project, and are for the benefit of subcontractors and suppliers. The payment bond guarantees that subcontractors and/or suppliers will get paid if the General does not or cannot pay them. If the General does not provide this bond, and the Owner neglected to assure that the General had provided it, then the Owner can be held responsible for any payment if the General is no longer viable.

Claims on the Payment Bond

As we discussed earlier, subcontractors and suppliers that do not have a direct contract with the General must provide a Preliminary 20-Day Notice if they wish to assert their rights to a claim on the bond, just as they do for Stop Notice claims. A sub or supplier must also notify the surety that they are making a claim by filing a "Claim on the Bond" that lays out, similar to the Stop Notice form, the amount that is due and owing, and the reasons why it is due and owing. The contact information for the bond surety is, or should be, provided by the General, and must be made available to all subcontractors and suppliers.

If a claim is made on the payment bond, the surety will request information and documents related to and supporting the claim. The surety will also pass the information on to the General and ask for the General's response. The Owner is not involved in this type of claim, and there is no requirement that the Owner hold any funds. The bond is a completely separate source of funding for subs and suppliers that are not paid.

Perfecting the Bond Claim

A claim on the bond must be perfected by a lawsuit. The limitation period for commencing the action is 6 months after the time when a Stop Notice can be filed. Assuming the Owner properly records a Notice of Completion (or Notice of Cessation), any Stop Notice must be filed within 30 days from the date of the recorded Notice of Completion. The lawsuit to perfect the claim on the payment bond must therefore be filed approximately 7 months from the date of the Notice of Completion; 30 days for the Stop Notice, plus six months.

On the other hand, if there is no Notice of Completion or an acceptance, you will have 90 days after a cessation of work, plus 6 months, to file a lawsuit on the bond. These time limits are only important if you have missed the time frame to file a Stop Notice. Otherwise you will be filing your lawsuit on the Stop Notice and will include the bond claim as well. There is no reason to do it any other way.

Attorney Fees in Payment Bond Claims

In any action on a Payment Bond, by statute, the prevailing party gets its attorney fees. This is so even in those instances where the General and the sub have a provision in their subcontract agreement that each side will bear their own attorney fees in any litigation or dispute.

If the sub and the General resolve their dispute at arbitration or mediation, and that dispute includes a claim on the bond, then unless the Settlement Agreement or judgment includes a provision that all fees and costs for both the surety and the General are settled, the sub is free to go after the surety for the attorney fees, and it will likely prevail. (21)

This is an important issue for both subs and Generals to consider. Subcontractors can backdoor a fee claim against the General, notwithstanding the express terms of the Settlement Agreement, the arbitration award, or the subcontract that may preclude fees. If the Surety is not released as part of the settlement or resolution of the case, and is then required to pay fees or any other costs, the General, as the Principal on the Bond, or the personal indemnitors, will be required to reimburse the Surety. It is important to make sure that the Surety is nominally a party to any arbitration or mediation and that it is released by the resolution of the dispute with the General.

It's a better practice to include the release language for the Surety in the Settlement Agreement, and better still to have the Surety participate as a party in any arbitration or litigation. The General will be present, the Surety has all the same defenses and will be defended by the attorney for the General, and there is a unity of interests between Surety and General that makes it logical for the Surety to "participate" in any proceeding. Any decision or settlement, therefore, should automatically include the Surety.

Notice of Completion

The Owner is required to record the Notice of Completion within 10 days of completion. If it doesn't, then it is invalid if recorded later. The determination of when the project is complete is left to the Owner's discretion, notwithstanding that it is also a defined term in the Code. By Code, Completion is defined as actual use, or beneficial occupancy, but the Owner gets to apply the definition when and how it wants. It is not uncommon for the Owner to insert a provision in the contract that provides that its use of the facility does not mean that it is accepted or complete. How the facility is used and what remains to be done however should determine the status of completion, not the Owner's self-serving definition.

The Owner can record the NOC after completion and, while it is technically invalid, it may have the effect of re-opening deadlines that were presumably closed; i.e. for Stop Notice claims and bond claims. Recall too that the NOC on a public project triggers the release of retention within 60 days, which explains why the Owner might not want to record it any earlier than it absolutely must.

The Owner is not required to tell the contractors that the Notice of Completion has been recorded, although most will. Subcontractors can pay $2 when they file a Stop Notice. If they do, the Owner is required to tell the subcontractor when the deadline to file a Stop Notice will run, which is contingent on the NOC being recorded, and so indirectly provides notice of the recording.

The Notice of Completion is an important document. Watch for it and calendar when it is recorded. It has significant consequences.

> **NOTE:** A late-recorded NOC can extend the time for the Owner to file a labor compliance action. By Code, the Owner or its Labor Compliance Program administrator have a total of 300 days from the recording of the Notice of Completion, or acceptance of the project, to file a complaint with the Director of Industrial Relations for forfeitures and/or penalties. If the Owner postpones recording the NOC, then it does not need to be timely on a number of other requirements, including claims with the DIR.

Termination

In public contracting, there are two types of termination. One of the terminations is disappointing, but perhaps not so bad. The other type of termination is bad, very bad, and is not something that you want to ever occur.

Termination for Convenience of the Owner

The not-so-bad termination is the termination for convenience of the Owner. Termination for convenience is a right that only the Owner has in public contracting; the General does not have this ability. Termination for convenience is when the Owner decides that the project is not worth continuing for one reason or another and issues the General a form of stop work order that will terminate the contract. The right to terminate for convenience, and the manner that termination for convenience occurs, will be described in the General Conditions.

If the Owner decides to terminate for its convenience, it will be obligated to pay the costs that the General and all the subs have incurred to that point in time, and any additional costs that cannot otherwise be avoided. In this situation, the Owner is responsible to pay all outstanding amounts that are due; i.e. progress payments, agreed to change orders, retention, that sort of thing. The Owner still has the right to negotiate change orders that may be outstanding, but it must resolve them or the contractor's right to make a claim comes into effect.

In this situation, the General and subs are entitled to a fair profit, even though they will not do the work. The value of the expected profit is a valid cost that the contractors are being forced to give up because the contract is being terminated. The contractors assumed and relied on the project for a certain percentage of their overhead and the premature termination by the Owner is depriving them of that revenue. This is a cost that the Owner should pay as well.

The Owner will scrutinize all costs very closely, so overreaching is not a good idea. The False Claims Act can be asserted if the contractors grossly overstate their entitlement to reimbursement. Good documentation and legitimate demands are expected, required, and necessary.

Transitioning Project Responsibilities

When the Owner directs you to stop work, do it. Of course, if there is an open trench that poses a safety risk, don't just leave it. Common sense is required, and if there is a logistical issue, what needs to be done to make the site safe and secure, it is imperative to ask the Owner what and how they wish the work to be closed up. Do this in writing and get directions in writing; verbal directives are OK, but confirm them in writing. You do not want to be responsible for leaving the site in an unsafe condition.

Liability for the site at the time of termination and the demobilization effort is something that needs to be set out in writing so that it is clear who is going to be responsible for the site and when that transition will occur. Remember, while the General has control of the site, the General, and through the various indemnity obligations, the subs, are liable for all damages, losses, and claims that occur while they are on site and perhaps even after they leave. Be clear who has the liability after the job is shut down, and when that liability is transferred. Make sure there is a clear transition of responsibility and that the transition is confirmed in writing.

When there is a termination for convenience, the Performance Bond and Payment Bond Surety need to be placed on notice. The Bonds need to be cancelled, and the Surety must be informed when this will occur.

Likewise, the insurance carriers, especially the Builder's Risk carrier, should be notified. Liability, and coverage for liability during the shutdown period and afterwards, needs to be resolved. In a termination for convenience there are a lot of housekeeping issues that need to be addressed, but it is usually a fairly amiable process that all parties have an interest in getting completed in an expeditious manner.

Termination for Cause

Termination for cause is the functional equivalent of the Owner deciding that the contractor is in material breach of the contract. The Owner is exercising its rights under the law and the contract to terminate the contract and throw the General off the project. This situation is the last straw, and it is not undertaken lightly by any public agency, which is not to say that the Owner is always right when it takes this action. As often as not, the Owner is not correct and the termination for cause is not legitimate.

Termination for cause is never a surprise; the project will be in trouble long before this type of drama is played out. In a termination for cause, the Owner and contractors have come to loggerheads, and the project is either significantly behind schedule or it is being completed in a manner that is contrary to the plans and specifications, or both. Termination for cause is a nasty process, and as the process gets closer to conclusion, the relations between the parties get more and more strained and contentious. Litigation is unavoidable, and it will be long and costly.

In order to exercise the option to terminate, the Owner must have good cause and it must follow certain prescribed steps. It is not simply a matter of showing that the project is delayed. The Owner must show that it advised the contractor of the delay, it directed the contractor to cure, it provided a reasonable time for the contractor to cure, and the contractor has either not cured or is incapable of curing the delay or the default. There is a lot to this requirement.

Before a termination for cause can issue, there must be action by the Board or Council, or whatever governing agency is in charge of the project. This will include a hearing where the contractor can defend itself and make its case that a termination is either unjustified or premature. As a practical matter, these hearings are more show than substance.

The Owner's representatives, whether they be the project manager, the attorneys, or some other representative, will have advised the Board or Council long before the contractor shows up to try and explain why it is not at fault. The Board or Council will be predisposed against the contractor before the contractor is even aware that termination is an option being considered. The Board or Council may give an extension of time to the contractor to cure the default, and this may be a genuine effort to be fair, but you can be sure that if the matter is before the Board to decide on termination, that any information provided to the Board is not favorable and that the intention is to get rid of the contractor, not keep them.

The Cost of Termination for Cause

If the termination for cause does issue, then the General is off the project. Progress payments that were due and owing will not be paid, change orders that were agreed to will not be paid, no further change orders will be negotiated, and retention will not be released. The Owner will hold all the funds until the project is completed and will use those funds to complete the project in any way that it wants.

Perfectly good work may be declared deficient and taken out so that a revised and reconsidered design can be implemented under the guise that the original work was not done properly and was defective. The General may not be able to review the work before it is taken out, but the Owner's experts will provide a report that will justify any and all actions taken. The Owner is not going to assist the contractor at all.

The cost of completion is not a major concern for the Owner. The Owner in a termination for cause will demand that the Performance Bond Surety take over the project and complete it. The penal sum of the bond is the contract price, and depending on the timing of the termination, there may be a fair amount of money available. The Owner also has all the contractor's funds, including any retention that was not paid prior to the actual termination. In short, money will not be a problem.

The Surety's Role In the Completion After Termination

When the Owner terminates the contractor, it will also tender the completion of the project to the Surety on the performance bond. If the Surety accepts the tender, it will need to hire a replacement contractor and complete the work. The Surety should have control of this selection process, but that is not always the case.

The logical choice for the replacement contractor is the General that just got terminated because, 1- that General knows the project best and 2- that General has a vested interest in getting the work done, in a timely manner, and for a reasonable cost, assuming that the General is still viable financially. Remember, the General is the Principal on the bond and so is responsible for all costs the Surety spends to complete.

Likewise, when the Surety completes the project, it is entitled to the amounts that remain on the original contractor's contract; all retention, progress payments, and change orders that were withheld from the terminated contractor. Under these circumstances, the terminated contractor is the best choice to complete the project because of its vested interest and desire to control costs.

The Owner will not accept the original contractor, especially if the reason for termination is poor or deficient work. The question then is- who has the last word on the selection of the replacement/completion contractor? If the Bond provides that the surety can choose the replacement contractor, then it should be the Surety. Therefore, the Surety and terminated General will need to find a suitable replacement contractor that will not gouge the Surety, that will do a proper job in completing the project, and that will be acceptable to the Owner. This is not always easy.

The Owner, if it is seriously considering termination for cause, will, or should, know whom it wants as a replacement contractor even before the termination is put in place. The Owner may enter negotiations with the Surety to get its preferred contractor, or it may insist that a certain contractor be used. The Owner has a lot of influence in this situation because the longer the Surety and Owner argue about replacement contractors, the more costly the delay becomes. Once the Surety accepts the Owner's tender, it must act reasonably and, if the Owner plays it right, the Surety can be put in a position so that it looks like it is either being unreasonable, or worse, breaching the agreement to complete.

Once the replacement contractor is selected, whether it is the Owner's preferred contractor or another contractor, the completion contract will be awarded. From then on, the surety and terminated General need to be especially vigilant. There is nothing to keep the replacement contractor from running up bills based on claimed deficiencies in the existing work and the need to replace that deficient work. It is easy to hide change order work that should not be paid by the surety/contractor by claiming it is not a change at all but the result of the terminated contractor's originally defective work.

In short, it's a mess, and by the time the project is finally completed, the terminated contractor or the Owner, or both, will have filed a claim and initiated litigation that will consume even more money and time. The subcontractors that have not been paid under the original contract will also have filed lawsuits.

Hold Agreements will be used to forestall any subcontractor litigation by agreeing to pay some part of the funds or making some other arrangement. For example, the original subs may be issued new contracts to complete the work in return for a payment of money and their agreement to return to the project. Their retention may not be released but certain outstanding change orders may be negotiated so that closure of some outstanding disputes will encourage their cooperation in the completion take-over agreement.

It is likely that the Surety and the terminated General will have their own issues resulting from the handling of the completion contract. If the stress of the project leading up to the termination was bad, the process following termination will be more than all of that by exponential factors.

It is not at all certain that the Owner will prevail in the following litigation, but remember that the Public Owner has an advantage that contractors do not: they can tax and raise money for the litigation, not to mention they have all of the terminated contractor's money that they can use to fund the litigation. They will spend that money, too. Even if mediation resolves the issues, the settlement will be substantially less than what the terminated contractor will have spent to defend itself, what it will need in order to reimburse to the Surety, and what it will have lost on the project.

Bottom line, termination for cause is a bad affair any way you look at it and is costly in every sense of the word.

Miscellaneous Public Works Issues

State Contracts

Contracts with the State are generally the same as contracts with local public agencies, and most of what has been discussed already will be applicable to both. There are certain things that are different, and you need to be aware of them if you are pursuing a State project.

For one thing, at the time of bid, the contractor has a statutory obligation to meet certain minority participation goals, including involving women and minority contractors in the bidding and construction process. The State will monitor these efforts as part of its review process, and bidders will need to submit a report on their efforts to satisfy these goals. Any willful or deliberate effort to circumvent this requirement by misrepresenting the outreach efforts can, if they are proven, result in significant penalties; up to $50,000 for the first offence and up to $200,000 for subsequent violations. The contractor will also be suspended from bidding on projects for a minimum of 30 days and for not more than a year for the first offense, and for subsequent offenses, up to 3 years.

This debarment (suspension) from bidding has significant repercussions that will follow the contractor for longer than the debarment period. As discussed earlier, on any public contracting project there is usually a pre-qualification process. If a contractor has been debarred by any agency, State, local, or UC, there is little chance it will be pre-qualified on any public project, regardless of the Owner.

If there is a claim for money due on the contract between the State and the contractor, a statutory Government Code section 910 claim is not required. Once you have filed your claim with the State, and after you receive the decision, the Public Contract Code specifies that arbitration is the dispute resolution method that must be used to resolve your claim. The contractor must comply with all contractual and statutory prerequisites before commencing arbitration, but once those are satisfied, arbitration, not litigation, is the required way to settle the dispute.

You must commence the arbitration within 90 days of the department's final determination on the claim, i.e. rejection. Commencement means making a demand for the arbitration, not necessarily beginning the arbitration itself. If the parties decide that they don't want to arbitrate, they can waive arbitration and proceed to court. Again, become familiar with your contract claims procedure and watch the time.

Retention must be withheld on any State contract. By statute, the minimum retention is 5% of the contract amount, but the State can agree to reduce that amount once the project is 95% complete. The General can substitute securities for retention, just as it can with public agencies. The General must also pay its subs in a timely manner. If it does not, then the same types of penalties and interest attach to the withheld payments as do for any other public agency.

> **Note:** One interesting and important aspect of the State Contract Act is the availability of payment for mobilization costs. For State contracts of more than $25M, the contractor can bill for 50% of its mobilization costs when 5% of the contract value is earned. There is a sliding scale arrangement that is provided by Code, and it describes how these payments are made and in what amount they are allowed. It's an opportunity for an upfront cash infusion that should not be overlooked or ignored. Make sure you have the appropriate line item on your Schedule of Values for mobilization, so you can take advantage of this opportunity to bill for these costs.

Claims Against the California University System & UC Regents

Contracting with the California University system is pretty much the same as contracting with the State or a local public agency. As with State contracts, there is no requirement for a Government Code Claim (the statutory 910 claim) for claims. There will be a claim procedure in the contract, and the Code prescribes a 6-month time limit for filing the action to recover on the claim, but you do not need to file a statutory Government claim before you file litigation.

> **Note:** There are two university systems for purposes of the Public Contract Code. There is the University of California (UC) system that is covered by Public Contract Code section starting at section 10500. There is also the California State University (CSU) system that is covered starting at section 10700. Be clear which University system you are dealing with. The procedures and requirements are essentially the same, but they are two different systems.

Lease-Lease Back & Design Build

One of the more curious aspects of public contracting is the way that the Legislature has created an enormous loophole for school construction. It is the Lease-Lease Back system that is provided at Education Code section 17406. Under this provision, a school district can avoid the requirements of the open bid/lowest responsible bidder procedures by using the alternative procedure in the Education Code.

Briefly, Lease-Lease Back works like this. The school district has a piece of land that it wants to build a school on, or an existing building that it wants to have refurbished. The school board votes to use the Lease-Lease Back procedure and thereafter invites certain pre-selected contractors to provide proposals for the project. The District may already have plans, or it may want the contractor to put together the design based on a performance outline of the type of school building that is desired; what type of materials, the energy requirements, any LEEDS requirements, whatever the Board wants.

These pre-selected contractors make their presentations to the Board, or to a panel designated by the Board. A contractor is selected and submits a guaranteed maximum price ("GMP") for the project. The GMP can be increased only if the District makes changes to the building program.

Any changes that are required in the plan in order to build the project are the responsibility of the contractors.

The Board then votes to accept and award the Lease-Lease Back contract. Once the contract is awarded, the District may go to court for a validation hearing. At the hearing the District asks the court to approve (validate) the contract and if the court does, no one can challenge the contract once the project is underway.

The Lease-Lease Back agreement usually contains a number of separate agreements. There will be a site lease where the contractor leases the project site from the District for a nominal fee, usually $1. There will also be a facilities lease agreement, the Agreement (contract) between the school district and the contractor, and a set of General Conditions that are usually modified General Conditions from the competitive bid projects that the District has done in the past.

> **NOTE:** Be especially vigilant in reviewing these various documents. Sometimes there are inconsistencies and those inconsistencies can cause enormous problems. Also, be very careful about what other Codes may be included by reference in the Lease-Lease Back. If the Public Contract Code is identified, be sure to limit its impact by identifying what statute is, or is not, included. You do not want provisions that relate to competitive bidding to apply to the Lease-Lease Back and thereby cause confusion about how the Lease-Lease Back agreement will be governed. This is an alternative procedure to the competitive bid so you don't want competitive bid protocols to apply.

The contractor might be expected to fund the project by arranging financing and proceeding with the construction, or the District may use its own bond funds to pay for the work. The monthly progress billings are submitted to the District, and are paid as lease payments. As the billings are paid, the lease is bought out, so that by the end of the construction, or shortly thereafter, the "lease" is paid off and the school district "owns" the building and the property. For cash-strapped Districts it is a great system, but some fairly well off Districts also uses it so that a project can be handed off to a favored contractor who then builds it for the District. Nice work when you can get it.

Design-Build: Another alternative way for public agencies to avoid the open bidding system is the design-build approach based on best value analysis. This project delivery system is permitted for projects with a value in excess of $2.5M, and there is a fairly comprehensive statutory scheme that must be satisfied in order to implement design-build. Basically, the local agency reviews the proposals and chooses the one that gives it the best value for the type of program that is desired. The process is not necessarily consistent with the purpose of the open bidding rules, but under the circumstances, and if used properly, it provides the local jurisdictions with a way to get more for the public dollars. For more on this sort of construction arrangement, refer to Public Contract Code section 20133.

Conclusion

Public works contracting is not simple, in fact, it is quite complicated. If you are not familiar with the requirements of the Public Contract Code or the Government Code, there is a very real possibility that you will run into some very serious issues, especially payment issues. It doesn't matter how good a builder you are or how well you performed on the project. If you do not follow the rules, you do not need to be paid. The Public Contract Code and the Government Code are very sophisticated statutes that are strictly applied by the courts. If you are new to public works contracting, be sure you understand what you are getting into. If you have only done private works contracting, be aware that it is not the same as public works contracting; not by a long shot.

Even if you are an experienced public works contractor, there are traps that can befall you. It's important that you read the contracts before you sign them. Make sure that you understand what is required of you for submitting payment applications, change orders, and especially claims. Make sure you know whether or not a Government Code Claim is required for any claims that you want to pursue at the end of the project. Make sure you know whether the contract claims procedure is exclusive or if there are other requirements. Finally, make sure you know who you are contracting with, and who the owner is, and if you have not dealt with the owner before, ask about them.

When you have done all that, good luck!

Chapter 8

Litigation

LITIGATION IS NOW A fact of life for the construction industry and, in a rather perverse sort of way, it is a rite of passage. Being involved in a lawsuit is indicative of your toughness and your willingness to fight. Just like the bullies in the school yard will take notice that you will fight back, even if you lose, and are therefore likely to leave you alone, contractors and Owners know who they should negotiate with and who they can push around in large part by their reputation for being willing to engage in litigation.

Understanding what is involved in the legal process is important to evaluating when to fight and when to roll. Not every dispute is worth the cost, and discretion, as they say, is sometimes the better part of valor. The other part of a winning strategy is planning ahead and documenting your projects. The more you can put on the table that proves your point, or at least demonstrates that you have a point, the more likely the other side is to take note and to reconsider their own position.

At a negotiation, the side that has a comprehensive file that documents the way a dispute has developed, and why it is now an issue, is likely to be more persuasive than the side that just makes a lot of noise. If you can lay it on the table, you're going to get the other side's attention. If you cannot, then it should get your attention that maybe you do not want to fight about the particular issue, and therefore whatever you can get is what you will take, half a loaf being better than none.

There are times when a lawsuit and litigation are inevitable. The time-frames are often predetermined, because if you do not file the lawsuit you will exceed the statute of limitations and lose the right to bring the claim. We discussed, for example, Mechanic Lien actions, Stop Notice actions, and Bond actions, all of which have particular limitation periods to file a lawsuit. If you exceed those time limits, you lose the right to make that claim.

Likewise, and with even more dire consequences, are the time limits and prerequisites for filing actions against governmental agencies. If you do not conform to the Codes and/or the contract requirements on public works projects, you will lose all your rights, and there are no second chances. Once you're out of court because you did not follow the Code or contract, you're done, short of any reversal on appeal; a very costly affair in and of itself.

Satisfying the Prerequisites to Litigation

Before you can get into court, there are certain critical prerequisites that need to be met. The contract and the type of project, public or private works, will determine what and how those prerequisites are addressed.

If the project is a public works project, you need to be concerned about both statutory and contractual prerequisites. The contract with the public entity, and the type of public entity that you are doing business with, will determine how you must proceed. If the public entity is a local public entity, and the contract provides that the contract claims provision is the "exclusive" means to settle any dispute, then that is the way you settle the dispute. If the contract requires as part of that exclusive procedure that you also provide a statutory claim, then you must provide a Government Code Claim pursuant to Government Code sections 900 et seq. If the contract is silent about exclusivity, then you must provide a statutory claim.

> Note: Recall from the Public Works section: If your contract is with the local public agency, you must provide a Government Code Claim. If the contract is with the State or with the UC Regents, you do not need to file this statutory claim. There may be other contractual requirements, but no formal Government Code Claim is required.

If you must provide the statutory claim, you should have an attorney, not because it is required, but because it makes sense. This is a very complicated area of the law, and you need good counsel. Even if statutory requirements are not specifically required, you should have an attorney review the contractual claim procedure. It is money well spent.

In private works, the failure to adhere strictly to the letter of the contract does not have such draconian effect. However, there are situations where the fact that you, as a subcontractor, or even a General Contractor, did not send a Preliminary Notice, or failed to provide an Owner notice of a Mechanic Lien, or failed to file a Stop Notice or record a lien can have significant consequences. You can still bring an action to recover money, but you will have lost the right to assert some very important rights.

The point here is to call to your attention to all of what we discussed previously regarding what rights you have in various circumstances, and what you must do to preserve those rights. If you don't do it properly, you will lose those rights, and sometimes you can lose them forever. Pay particular attention to your contract terms and conditions, so that if you actually need to pursue your rights in court, you can.

The Lawsuit

First, a short primer on the psychology of the lawsuit. If this is the first time that you are being sued, you will run the full gamut of emotions. First, you will be outraged, and likely angry, next retaliatory, and want to get back at the *sumbitch* that sued you, then you will want to fight it, then you will grow tired of it when it drags on and on and on, and then you will get worried, grow indifferent and want it to go away, at which time you will likely enter mediation, where, if the matter settles, you will come away disappointed and feeling as if you got a raw deal and paid too much, or, conversely, you didn't get paid enough. If the dispute ends up in trial, there are a whole different range of emotions.

In short, lawsuits sound great in theory, and you will be buoyed by the attorney's assurance of the solidity of your case, but after a while you will grow weary of it all, of waking up at night thinking about it, of the time it is taking out of your life, of the distraction and inconvenience, and ultimately of the unfair result that you will perceive is reached by any settlement. You are not alone, but that is the nature of the beast. As they say, be careful what you ask for; you just might get it.

If you sue someone, you are the plaintiff. If you are sued, you are the defendant. If you are a defendant but then sue another party, then you are the cross complainant and the other party is the cross defendant. There can be any number of plaintiffs, defendants, cross complainants, and cross defendants. If you do not know the names of all the defendants or cross defendants at the beginning of the lawsuit, and you want to preserve your right to bring them in later after you get their names, you name them as Does, Moes, Roes etc; usually some word that rhymes in that fashion.

A defendant must typically answer within 30 days of being served with the lawsuit. If you are a defendant and you have a claim that is related to the same transaction that gave rise to the Plaintiff's lawsuit, you must file what is called a mandatory cross complaint at the same time as you file your Answer to the Complaint. The cross complaint is then served on the cross defendants, which can also include the Plaintiff as a cross defendant, and any other parties you name in the cross complaint. These cross defendants have to file an answer, and these cross defendants have the right to file their own cross complaints.

> **Note:** There are also permissive cross complaints. A permissive cross complaint is one that is not directly related to the same transaction (project), but that would still involve the Plaintiff. For example, if you are sued on one project but you have a claim against the same Party for another project, you can bring it as a permissive cross complaint if it will not unduly interfere with the case you just got sued in. If it does, you can always bring it as a separate action, and you would be the plaintiff and the other guy would be the defendant.

It is all very logical in a bizarre sort of way, but the end result is that all the parties and all the claims of all the parties related to the underlying project are brought together in one lawsuit. At least, that's one way that it can happen.

Sometimes, especially on public works projects, the subs will each bring individual actions against the General. These lawsuits may involve claims that should go against the Owner, and so the General will sue the Owner and allege causes of action related to the subcontractors' claims. In this situation, the court, rather than having a bunch of lawsuits all related to the same project and all against the same General, will consolidate the cases. This makes it easier for the court to manage the caseload, but it also means that your lawsuit is just one of many. Depending on your claims, this can be good or bad, but that's the way it will be handled.

Complex Case Designation

Cases are designated complex or not complex. A complex case is one where there are numerous parties and/or the issues involved are complicated and likely to involve complex legal issues. The effect of the complex/not complex designation is the time allowed to get the matter to trial. Complex designation means that the case is taken off of the fast track trial calendar that all California courts are required to comply with. Non-complex cases are expected to be brought to trial within a year of filing or becoming "at issue," which means that all the pleadings are filed and everyone that needs to be in the case is in the case.

Complex cases do not need to be resolved in the one-year timeframe, and they rarely are. They can go on for years, as long as some part of the case is brought to trial within 5 years of the date the case was first filed. The longer they go on, the more costly they become and the more difficult to resolve.

Limited and Unlimited Jurisdiction

If the amount in dispute is less than $25,000, the matter is a limited jurisdiction matter, and different rules apply to the way that the case is handled and what the parties can do in preparing for trial. Over $25,000 is unlimited jurisdiction, and this is typically where construction cases end up. The type and amount of discovery allowed is far more expansive in unlimited jurisdiction matters, and no jury is permitted in a limited jurisdiction matter.

If the case is less than $10,000 in value, it is a small claims matter, and should be resolved in that forum if for no other reason than that it's faster and a lot less expensive. No attorneys can represent a party in small claims court, and the decision for the plaintiff is final. Only the defendant in a small claims action can appeal the decision to the Superior Court, and if there is such an appeal, attorneys can be present for the appeal. Given the amount at issue, appeals are not usually taken, but if you get a bad decision from the judge pro tem (a volunteer judge) then it may be worthwhile to have a more experienced judge hear the matter.

The Complaint

The Complaint is comprised of a series of general allegations, or facts, that serve as the basis for more specific allegations that are listed under the various causes of action alleged. The causes of action are numbered, for convenience when identifying them, and not in any order of importance. It makes no difference how they are set out in the Complaint, because each one needs to be a stand-alone claim against the other party. That is, each one of the causes of action needs to be supported by one or all of the general allegations and the facts that are alleged, and any of the facts that are incorporated by reference into the individual cause of action. Attorneys usually try to make the Complaint read in a way that makes some organizational and logical sense, and will typically include the more important claims first.

An interesting, but underappreciated, point is that the facts alleged in the Complaint do not have to be accurate or even provable in order to state the cause of action. All that needs to be alleged is information sufficient to "state the cause of action," and the lawsuit is launched. Stating a cause of action is relatively simple, given that for any cause of action that is alleged there are specific elements that need to be stated.

For example, in a breach of contract cause of action, the basic allegations that need to be alleged are: 1- there is a contract; 2- it is written, oral, or implied; 3- the other side breached the agreement; 4- the plaintiff performed everything it was supposed to perform, except what was excused by the other side's breach; and 5- that there were damages to the plaintiff as a result. The above recitation would satisfy the pleading requirement for a claim of breach of contract without much more factual information. Most complaints will go into detail regarding the project, the nature of the contractual relationship, and how the defendant breached the agreement.

This is not to say that anything goes when you're preparing a complaint, or when you receive a complaint. The law does not permit deliberately outrageous or extreme statements that are only intended to embarrass the other side. If that happens, there can be a Motion to Strike or a Special Demurrer that will attack those types of pleadings.

More on those terms later, but suffice it to say that the Complaint can pretty much say anything, provided the specific elements of the cause of action are satisfied and there are "facts" to support the allegation. It doesn't matter whether you can prove the facts at the pleading stage; all you need to do is satisfy the elements of the cause of action. Therefore, when you read the Complaint, especially if it is your first time being sued, you need to read the claims with a bit of a jaundiced eye, because a lot of what is stated is likely exaggeration or just plain bullshit. Don't take it personally!

Important Pleading Allegations in the Complaint

There are certain claims that you do need to take very seriously, and certain pleading requirements that you must satisfy.

1. If the Plaintiff alleges in the general allegations that it is a licensed contractor permitted to do work in the state of California, which is a more or less a required allegation, and the defendant Answers by making a General Denial of all the plaintiffs facts, then licensing is placed at issue. This means that you, the Plaintiff or Cross complainant, need to obtain from the California Contractors State License Board a certified record of your license for whatever time period is at issue in the Complaint or cross complaint. You need to provide a copy of the certified record to the other side when it comes time to exchange documents, and if not then, you need to have it ready for trial. If you don't have that proof, it could be that you will not be able to establish that you were properly licensed at all times during the project and therefore have no rights to assert in court. This can be a tricky issue and one that is easily overlooked, but it is something that is easy to resolve.

2. If you are suing a public agency, you need to allege that you have satisfied the claims presentation requirements of the Code and/or the contract, or that you are otherwise excused from compliance. If you do not include this allegation, then the Complaint can be attacked by the public agency and you will need to amend the complaint to include this allegation. Amending a complaint is not especially difficult, but it does delay the litigation, and if there is an attorney involved, and there should be, it is going to cost something.

3. If your company is a corporation, you must have an attorney to represent it. Even if the corporation is just you, and no one else is involved, you need an attorney to represent you in court. Something to keep in mind regarding corporations, especially small corporations with one shareholder, director, and officer, is that you can be sued as an individual, notwithstanding the corporate status.

   If you contract as a corporation, but you personally perform the work, the corporate shield may not protect you from liability, and you, personally, can be liable. In one case, the corporate veil was pierced and the court and held the president liable because he worked on the project and the work he did was done negligently and caused damage. The corporate shield protects corporate decisions (i.e. which projects to bid on or what to bid for them), but if you also do the work, the shield will not protect you if you do it poorly.

4. If a Release Bond for a Mechanic's Lien or a Stop Notice has been posted, then the surety issuing the bond is the correct party in the lawsuit, not the Owner and/or the lender. For example, if it is a public project and a Stop Notice has been filed, and the General has bonded the Stop Notice, the Public Owner is not a proper party in that litigation; the bonding company is. Naming the Owner could result in a false claim act violation.

   Similarly, if a Mechanic Lien has been recorded and a Mechanic Lien Release Bond is posted, the owner of the property is not a proper party, the surety is. If a lis pendens was recorded for the Mechanic Lien, it must be released as well, if a bond is posted. There can be some fairly serious consequences if you do not name the proper party, especially if you know a bond is in place and if you do not release the lis pendens. The property owner can bring an action against you for slander of title if you refuse to remove a stale or invalid lien and the lis pendens interferes with the owner's rights or ability to sell or otherwise finance his property. Otherwise, Mechanic's Liens are privileged and may be asserted even if they are ultimately found to be without merit.

Affirmative Defenses

We discussed the General Denial in an Answer, and how that will put at issue everything that is alleged in the Complaint. The Answer will also contain Affirmative Defenses, which are specific claims that are made and that introduce specific "new matter" into the dispute. These affirmative defenses will either refute the legitimacy of the allegations in the Complaint, and/or excuse the defendant's actions under one legal theory or another. The Affirmative Defenses do not address each and every cause of action but rather apply more generally and raise issues that apply to the Complaint as a whole.

For example, there are Affirmative Defenses that allege that the complaint fails to state a cause of action against the defendant; that allege that all claims in the complaint are barred by the statute of limitations; that allege that the damages suffered are the result of other's actions, not those of the defendants; or that the plaintiff is responsible or has failed to mitigate its damages; or that the damages alleged are offset by the damages that the plaintiff caused to the defendant; or that the Plaintiff is in material breach of the contract.

Just like the Complaint, the Answer, especially the Affirmative Defenses, can allege pretty much anything. In construction, there are standard Affirmative Defenses that are typically included, and there are others that can be included that are individual to the particular lawsuit. Just like Complaints, Answers can be amended, so if you need another Affirmative Defense, it can be added later if you get the court's permission.

Affirmative defenses do not make affirmative claims, however. Affirmative defenses can allege offsets, which are amounts that will be subtracted from any recovery by the Plaintiff. If the Plaintiff gets nothing, there are no offsets; even if you prove the offset amount, you will not recover anything. If you want money from the other side to pay for your damages, you must file a cross complaint.

Causes of Action

There are some fairly standard causes of action that will typically be alleged in any construction case. The following are some of those, with a brief description of what exactly is being asserted in the claim.

Breach of Contract Obvious, and no need to go into great detail; it is alleged that you had a contract, you did not perform the terms of the contract, you had no excuse for the non-performance, and the other side was damaged as a result.

Negligence This is a little tricky. There are generally two types of claims in any lawsuit; contract claims and tort claims. Contract claims rely on an agreement. Tort claims are for personal injury or damage to personal property, and the claim is based on standards that we all, as a civilized society, agree to abide by in order to enjoy our lives. It is the reasonable person standard; i.e. what would a reasonable person do in a similar circumstance?

In construction disputes, you cannot allege that someone was negligent simply because they performed the contract in a negligent manner. The courts have ruled this is a contract issue, not a tort issue. Your damages are contract damages, not tort damages.

However, if the work was done negligently and caused damage, construction defect claims include this sort of cause of action, then negligence is an available cause of action. Conversely, if the claim is intentional infliction of emotional distress because the contract was not performed properly, then there needs to be a separate tortuous act that gives rise to the claim. The failure to perform the contract properly is not enough.

Fraud This is a popular claim, but one that has very specific pleading requirements. In order to allege fraud, you have to specifically say what the fraudulent conduct was, and how you relied on that conduct to your detriment. Fraud is a tort-based claim, and it allows the plaintiff to seek punitive/exemplary damages. It is a tough claim to prove in any circumstance, and it is no different in the construction context. However, if the behavior of the defendant is so outrageous and/or malicious, the court could award exemplary/ punitive damages. Depending on the circumstances, these can be substantial.

Quantum Meruit This claim basically asserts that you want to be paid the fair value for your work, regardless of the contracted amount. Under quantum meruit theory, the contract is deemed not to exist for one reason or another, and therefore you want to be paid what it cost you to do the work. Quantum meruit cannot be alleged against a public agency, although a subcontractor on a public contract can certainly allege a cause of action for quantum meruit against the General. On private work projects, there is no such restriction and the Owner is fair game.

Account Stated Another fairly common claim that essentially says that the parties have reached an agreement regarding the amount in dispute, the plaintiff has not been paid that amount, and the defendant should be required to pay that amount. Under certain circumstances there can be an award of limited attorney fees but otherwise it is a fairly banal allegation and accomplishes little, except that it gives the Complaint more weight, mostly because of the extra paper involved when you include it.

Stop Notice, Mechanic Lien, & Bond Claims We have discussed these claims. The allegations for each are essentially the same; the plaintiff is one of the enumerated class of claimants, they filed a 20-Day Notice, they did the work, they recorded or filed the appropriate document (Mechanic Lien or Stop Notice), and they are now entitled to foreclose on the liens or stop notices or bonds so that they can be paid. These types of claims must be brought in the appropriate timeframe, and are usually part of the core claims in any construction dispute. The Mechanic Lien action can force the sale of the property and the Stop Notice action and Bond claims can generate attorney fees and interest.

Indemnity, Contribution, and Apportionment These are three similar types of claims that are typically found in a cross complaint rather than in the Complaint. Each of these causes of action seeks to have some or all of the other defendants pay for some or all of the damages that may be assessed against the cross complainant.

For example, there may be an allegation for express indemnity that seeks to enforce the indemnity provision included in the contract. It will state that there is an indemnity provision in the contract, and/or quote the provision verbatim. Depending on the type of project and the type of express indemnity included, there may be defenses to these types of allegations.

Alternatively, if there is no express indemnity provision in the contract, there may be a cause of action for implied contractual indemnity, or, more broadly, equitable indemnity. Implied contractual indemnity causes of action are equitable in nature; that is, they seek a fairness determination. The party alleging this claim wants the court to find that it did nothing wrong and therefore the other parties should pay all damages and should also pay it for its time and trouble defending the action, i.e. fees and costs.

If there is an express indemnity provision in the contract, then implied contractual indemnity is not available. You can't have a judgment for both express and implied indemnity; you get one or the other. It's not improper to allege both causes of action though because if the express indemnity provision is invalid, you still have the equitable claim.

Contribution is a claim that if you are found liable to the Plaintiff, then others should contribute to the judgment paid to the Plaintiff.

Apportionment is a cause of action that requests that the court make a determination of how much each party should be obligated to pay. It is based on the jury or judge's determination of what percentage of the bad acts were done by the various defendants, and that percentage is then used to allocate the amount of damages to each of the various defendants. If you are found to be 50% liable, you pay 50% of the damages, provided of course that the other defendants can make up the other 50% of the judgment. If not, then you might be stuck with the entire judgment, depending on the other facts.

Contractors State License Board Rules Violations

Be aware that the CSLB has a number of rules and regulations that can be the basis for alleged statutory violations. It is not clear whether these types of claims can only be raised before an administrative law judge that the CSLB would have, or if a trial judge or jury can also determine them. Basically, these are filler type claims. If there is an issue that is CSLB-based, you should notify the Board and they will take the issue up and investigate it. Bringing these claims in a lawsuit is a waste of time.

Contractor's License Bond Claim

Under the CSLB rules, if a contractor violates the rules and causes certain losses to another party, whether it is a homeowner or a contractor, the License Bond is there to pay for a limited amount of that damage or liability. License Bond sureties are not easily parted with their bond funds. There is usually a requirement that the demanding party prove that the offending contractor is actually in violation of the Rules. Even then, the surety is likely to contest any obligation that it may have, including that there was ever any violation by the defendant contractor; plainly, pursuing the license bond will take time and cost money. Any recovery is very limited, if there is any recovery at all, but it is an available fund if you want to pursue it; think small claims court if the facts are right and the defendant contractor is out of money.

The Lawsuit—File in the Correct Court

If you are the Plaintiff, the first thing that you do after you prepare the Complaint is file it at the Clerk's office in the Superior Court. The location of the Superior Court is critically important, and is dependent on where some, or all, of the defendants are located and/or where the project is located. The Plaintiff does not get to choose the location where the case will be heard, and if the matter is filed in the wrong court, the defendants can challenge the venue and force the Plaintiff to dismiss and re-file the action in the proper court, or else transfer the matter to the correct court.

A couple of things to consider if you are the Plaintiff and you have filed in the wrong court.

Transferring the Action

If you "transfer" the case, or agree to allow it to be transferred, there are additional fees that you will need to pay. You also need to get court approval for the transfer. The court that transfers the case and the court that will be receiving the case will both need to approve. This is a fairly standard procedure, but sometimes it is a bit messy and confusing, not to mention time-consuming and costly.

Dismissing the Action & Re-filing

If you "dismiss" the Complaint and re-file the case in the correct court, it is easier as a practical matter, but you also need to be careful about how you dismiss the action.

First, when you dismiss the case, you must dismiss it without prejudice. This means that you are not giving up your rights in the case, but are dropping the issues for the time being. If you dismiss with prejudice, you are basically saying you have given up all of your claims and the matter is now concluded. Unless this is what you intend, make sure that the dismissal is without prejudice so you can re-file your case.

Second, and very important, if you are dismissing simply to get to the correct court, you need to be very aware that by doing so any statute of limitations is running from its beginning. If too much time has elapsed since you filed your action, you may have run out the limitations period.

For example, assume you have a properly filed Stop Notice and the Notice of Completion is recorded. You filed the Stop Notice within the 30 days and you commenced the action within the 90-day time frame. For purposes of discussion here, assume you file your lawsuit on the 89th day, or one day before the last day to file the Stop Notice action.

After you file, you realize that the case is filed in the wrong court and you decide to dismiss without prejudice and then re-file in the correct court. You dismiss the case and then re-file it three days later in the correct court. You have just blown the statute. The other side will challenge the Stop Notice cause of action, and they will win. You had to file the Stop Notice action in the 90-day timeframe, but you filed it 92 days later. If you dismiss the case, the statute of limitations runs from the day it originally started to run and it is not tolled (stopped) while you had the case filed in the wrong court. Once you dismiss it, the clock runs as if it had been running from the beginning. (23)

In a more likely situation, assume you file your action timely, and then you serve the other side with the Complaint. The Complaint should be served within 60 days of its being filed, and the other side has 30 days to Answer. Assume further that you filed the Complaint on the 89th day after you filed your Stop Notice, you served the Complaint on the other side 30 days after the Complaint was filed, and then the attorney from the other side calls and tells you that it is in the wrong court. They call you two or three weeks after they get the Complaint.

You realize they are correct; you are in the wrong court. You then go to the correct court and file another complaint. You then go to the court where you first filed and dismiss that complaint. Have you protected the statutory time? You have not! You have again blown the statute. In both examples, you had to transfer the case, not dismiss and re-file it.

The point is that whether you are a Plaintiff or a Defendant, you need to look carefully at the court where the case is filed, and the Plaintiff, especially, needs to understand the ramifications of filing in the wrong court. It is possible to dismiss without prejudice and re-file, if there is time, but you need to be sure there is time to re-file if there are statute of limitations issues. If there are timing issues, transferring the case is your only option.

Statute of Limitations

Every cause of action has a statute of limitations attached to it. If you go beyond that limitation period, you do not have a right to bring the claim; for all intents and purposes you have no claim anymore. The statute of limitations commences to run when the cause of action has "accrued." This means that all the facts and elements of the cause of action are in place, and you can meet the requirements for the cause of action if you file a lawsuit.

For example, breach of written contract has a 4-year statute of limitations. Assume you are a contractor that has not been paid for certain work. The Owner has told you that he disputes his obligation to pay you anything more for the work that you did. When can you sue?

As a practical matter, you won't know whether or not the Owner is going to pay you until the end of the project. It's always possible that the Owner and you will negotiate certain issues and resolve the dispute before the project ends. In this situation, the statute of limitations would begin to run when it became clear that the Owner was not going to pay you for the work, which is likely when the project is complete, but the Owner does not pay you what is owed. You then have 4 years to file a lawsuit for breach of contract. You will undoubtedly file long before that time, but that is the outside limit.

Recall, too, that until the time for performance has come, there is no breach. If you have 9 months to get the project done, and completion of the project triggers the obligation to pay, then until the 9 months is up, your cause of action for breach has not accrued. If you know for certain that the other side will not pay you, you don't need to wait, but can bring the action when it is clear the other side will not perform or cannot perform.

Some of the other statutes of limitation that you should be aware of:

- Breach of oral or implied contract- 2 years;
- Negligence- 3 years;
- Fraud- one year from the time you are aware of the fraud, or if the fraud is the result of a contract, then 4 years;
- Payment bond claims on public works projects- within 6 months after the time to file Stop Notices;
- Payment bonds on private works- within 6 months, although it varies depending on whether it is recorded or if it is statutory or common law based;
- Stop Notice and Mechanic Lien foreclosure action- 90 days from the date of the recording (Mechanic Lien) or the filing (Stop Notice);
- Express indemnity- 4 years from the date the obligation arises. This is a bit tricky, because you can pursue indemnity claims after the principal lawsuit is concluded by filing another action for indemnity or you can pursue the indemnity claim in the principal action by cross complaint. It is obviously cheaper to put everything in one lawsuit, but there is the option to do it in separate lawsuits.

Serving the Lawsuit

In the usual scenario, a professional process server will come to your office and tell you that they have papers to serve. Those papers are the Complaint and the Summons, and you need to be served with both. There will be a number of other papers and documents from the court that must be provided. Don't lose any of them; they are important, and have important information.

The process server will hand you or your staff the documents and will ask you or your staff for their name and position in the company. Make sure your staff knows who can and who cannot accept service. If the only person in your office at the time is someone that cannot accept service, tell the process server to come back. Your staff does not have to accept service if they are not authorized to accept service.

When the process server gives you the documents, write down, or have the process server write down, the date and time of service at the bottom of the Summons. You need to know this because when you get served determines when you need to answer. The process server should write this information down, but more often than not, they forget. You should not; make sure you document this information.

There is an alternative manner of service, which is to put the Complaint and Summons, and all the other information, in the mail with a Notice of Acknowledgement of Receipt form. What this Notice says is that if the person receiving this Notice accepts this manner of service by signing and returning the Notice of Acknowledgement in the stamped envelope that is provided, that they will have 30 days from the date they sign and return the Notice to answer. If they do not sign it and a process server needs to be hired, then they may be required to pay for that process server.

If you receive this kind of Notice, and if you have an attorney, give it to him/her and let them handle it. If you don't have an attorney, you have to decide whether you will sign and return the Notice or blow it off and let them serve you directly. The Notice is a simple and fairly straightforward way to get service done. You are going to be served anyway, so this is an easy way to do it; no embarrassing visits from the process server and that sort of thing. On the other hand, why make it easy for the other side, especially if relations have deteriorated. Make them jump through the hoops if you want; you do not have to accept this sort of service.

One very important piece of information that you will find in the package, along with the Summons and Complaint, is the notice of the first Case Management Conference. It may be a separate document, or it may be stamped on the first page of the Complaint. The Case Management Conference is the first date that the court wants to see all the parties so they can explain to the judge what is going on and how the case is proceeding. It's typically some months off from the time that the Complaint is served. It is very important to calendar this date, and if you have an attorney they will need to receive this information along with the Complaint and Summons.

Insurance Issues & Litigation

Depending on the issues raised in the Complaint, there may be insurance concerns that need to be addressed. If the claims are that you did defective work and/or were negligent, and someone or something got damaged, you need to let the insurance company know immediately. This is where you tender the claim to the carrier and put them on notice that you have an issue.

Tender the Claim

The best and easiest way to tender the claim is by calling your insurance agent and telling them you got sued. Don't try and figure out for yourself whether it is a covered claim or not; let the carrier do that. Send your agent the lawsuit and ask them to process the claim. If you have an attorney, let them know, too, and let them know that you have sent the claim to the carrier. If you have an attorney who handles your legal affairs, give the Complaint to him/her and let them tender it to the carrier.

Insurance applies only to covered claims; that is, those claims that are not excluded by your policy. Reading an insurance policy is not recommended unless you have insomnia, in which case it is a sure cure. Knowing what is covered and not covered is a review process that either your agent or your attorney should undertake, but you should not. First, get the carrier's response; no sense wasting time figuring whether a claim is covered if the carrier is going to accept the claim. The carrier must respond to your tender, and you will know whether you have coverage or not when you get the coverage letter.

When you get the carrier's response, save yourself the aggravation of reading the whole thing. Read the first page or two and the last page or two. This is where you will find out if they are going to accept the tender and defend the claim. Assuming there is at least one claim alleged in the Complaint that would trigger coverage then, under California law, the carrier has to defend the entire lawsuit.

If there are covered and uncovered claims, or if there is some uncertainty whether the claims are covered, the carrier will send you a letter stating that they are defending the claim with a "reservation of rights." This means that they reserve all their rights to either change their mind later about the coverage if the facts change, and/or to seek reimbursement for costs incurred for defending claims that are not covered by the policy.

If the insurance company accepts the tender, they will appoint an attorney to defend your interests. Your personal attorney is not going to defend the case unless you pay for their time, or unless that attorney is also one of the panel counsels that the insurance company has, and that they send cases to when there is litigation.

Panel counsels run the gamut from great to adequate, which is one of the reasons for this book; so you can understand the issues, and so you can explain to any merely adequate attorney the issues in the lawsuit and why you are not responsible for the claimed damage. Likewise, you will be able to understand what the attorney is talking about when they call to discuss the case and coverage.

Insurer's Right to Settle the Case

One thing that you need to understand is that even if you are not responsible for the claims in the Complaint, and even if the claims against you are mere fluff and exaggeration, the carrier may chose to pay money to settle the case and be done with it. This is no reflection on you, your company, or your work. It is a practical reality of the insurance world, although it sometimes a really hard pill to swallow.

Understand that for insurance companies there are two pots of money that the carrier can pull from in a lawsuit; the indemnity pot and the defense pot. The indemnity pot is filled with the money that your policy has as the limits for any claim; i.e. $1M per occurrence, $2M aggregate. What this means is that when you have a covered claim, the carrier will pay up to the per occurrence limit to cover that damage or injury.

In most instances, the limits are far more than the amount that will be needed to pay for any claim. Most claims are settled through a mediation process, and the total amount of that settlement is funded by contributions from all the named defendants and their carriers. More on mediation in the next sections, but for now understand that the amount that the carrier has from the indemnity pot is dependent on your coverage limits, the amount of the claim, and the amount of potential liability you, the insured, has for the damage.

The defense pot is a different fund that the carrier uses to pay for the defense of the claim. This includes the attorney fees and any experts that might be needed to review the case and the alleged defects. If the contractor is a small player in the lawsuit, the amount of indemnity dollars is likely not very much. Regardless of the indemnity dollars, the defense dollars can run up very quickly. Lawsuits can drag on for years, and the attorney will need to participate in most, if not all, of the goings on in the litigation. This costs money, and these costs are paid from the defense pot.

It is not unusual for the carrier to pay minimal indemnity dollars, the nuisance value of the lawsuit, but still have significant defense dollar expenditure. Insurance companies, therefore, want to settle a case quickly, if they can, so they can avoid the costs of the attorneys and experts, especially if liability is clear and/or if liability is minimal. Therefore, you should not be insulted or angered if the carrier appears to "cave" to a demand in a case where you are certain there is no reason to pay anything.

Deductibles and Self-Insured Retentions

Insurance premiums are fairly pricey. One way to keep the policy's cost down is to have a deductible or to use a self-insured retention along with your policy.

A deductible will reduce your premium costs, depending on the size of the deductible; the bigger the deductible the better the reduction in rate. If you have a good run loss ratio because you do not have a lot of claims, then a high deductible is a good idea. Conversely if you have a lot of claims, then a high deductible is not worthwhile. You will get charged the deductible on all claims that are filed and that the carrier has to defend, regardless of whether you pay any indemnity dollars.

> **Note:** If the carrier settles the case for you, but then pursues another party for indemnity on that claim, or your deductible should be included in that further recovery and should not be charged to you. Whatever is recovered in the follow-on litigation should make both the carrier and you whole, and that means you don't need to pay out the deductible.

Self-insured retention (SIR) is where you agree to pay the first dollars in any litigation, up to a pre-agreed amount, and then the carrier picks up the rest. Obviously, the higher the SIR, the more you will pay. The savings should be correspondingly high in your premiums to make it worthwhile to have the SIR. The SIR is also on a per-occurrence basis, so if you get sued a lot in one year, the SIR gets paid out each time. Unlike the deductible, you will pay the first dollars, and until you do, the carrier has no obligation to pay anything.

Additional Insured Endorsements

Additional Insured Endorsements ("AI's") are typically required from subcontractors by Generals and Owners, as part of the subcontract insurance requirement. These endorsements basically provide that the additional insured entity, the General and/or the Owner, is provided the same coverage as the named insured, the subcontractor, and whatever claims are covered by the subcontractor's primary policy will be covered for the additional insured entity. The insurance requirement will typically require that the subcontractor's policy will be exhausted first, and that until and unless that policy limit is exhausted, the additional insured's policy will not be touched at all.

There are a couple of issues to be aware of with AI's. First, recall that in residential construction a subcontract cannot require a Type I or Type II indemnity. The subcontractor can only be required to indemnify for its own negligent acts, not everyone else's, and not the General's or Owner's negligent acts. The AI does not make that distinction. The AI makes the subcontractor's policy the primary policy, assuming the AI endorsement has 'primary wording' built into it, which is not always the case. Consequently (for the General), it is important that the endorsement clearly states that the AI coverage is 'primary and non-contributory' to clarify who will pick up the coverage if there is a question as to who is at fault for a given claim.

Second, if the subcontractor has an SIR, and if the SIR requires that the subcontractor, and only the subcontractor, can pay that SIR, then until and unless the subcontractor makes that payment, there is no primary policy at all, including no AI coverage.

The AI endorsement often (if not always) requires that there is a written contract in place before the form will apply. If you are the General, NEVER start a job without a written subcontract if you intend to seek indemnification (protection) from your subcontractors in the event of a claim.

Finally, make sure your sub's AI endorsement(s) apply to both 'ongoing' and 'completed' operations. Many AI endorsements will exclude coverage for 'completed' operations, which means that the general contractor will have no protection from their subs in the event of a construction defect claim. Most insurance companies will not allow 'completed operations' AI status for residential projects, so be aware that limitations may apply. However, 'completed operations' AI status is available and should be required by the General Contractor in any commercial construction project.

If you, as the General, have an AI with your subcontractors (and you should), and if you are brought into a lawsuit, you need to tender to the AI carrier as well as to your own carrier, or make sure your agent and attorney make the tender. Be aware, insurance companies are usually "reluctant" to respond to AI tenders, so don't be surprised if they don't pick yours up. The policy is there, it is subject to the same claims handling requirements that your primary insurance is subject to, and the bad faith issues remain the same for the AI carrier as for any other carrier. It is worth pursuing any and all AIs.

Cumis Counsel

There are times when the insurance carrier's interests can be adverse or in conflict with the insured's interests. If the coverage issue can be determined by the manner in which the case is presented to the jury, so that there is a potential that the carrier might not have a coverage obligation, depending on how those facts are set out, then the insured is entitled to *Cumis* counsel

Cumis counsel is another attorney than the one provided by the carrier to represent you. Under this situation, the carrier must pay for the other attorney, usually your own, to co-represent your interests in the litigation. It is a difficult set of facts that leads to the appointment of *Cumis*. Even if the carrier accepts your tender with a reservation of rights, that will not be sufficient in and of itself to entitle you to *Cumis*. However, if the facts are such that the way an attorney presents the case will decide whether there is coverage, and especially if the potential liability is significant, then you want to demand *Cumis*.

Let's assume that you are on a project where someone got seriously hurt. You get sued, and the claim is that you intentionally injured the other person or, alternatively, that it was inadvertent, and while you may have been negligent, you did not intend to hurt the other person. If the facts presented are that you and the other person had issues, and that you swung the 2 × 4 at the other guy, then that is intentional. If the fact presented is that you were holding a 2 × 4 and turned, and when you did the 2 × 4 hit the other fellow, then that is negligence. There is no coverage for intentional acts, but there is for negligent acts. How the case is presented will determine whether the carrier has to pay or not, and you need *Cumis* counsel to protect yourself in that situation.

Policy Limit Demands, Insureds Contribution, and Buss Motions

In some cases, the Plaintiff may make what is called a policy limits demand. If the policy has a $1M occurrence limit, the Plaintiff is demanding that amount as settlement of the case. This presents an issue for you and for the carrier. If the potential damages are clearly not valued that high, the carrier will likely reject the demand. However, if the case could be worth that amount or more, a rejection can become problematic.

If you are potentially exposed to a judgment that would exceed your policy limits, and therefore the policy limit demand is reasonable under the circumstances, you, as the insured, should demand that the carrier pay the limits and settle the case. It's entirely up to the carrier whether they will pay or not, but if they do not, and if the judgment at trial exceeds the policy limits so that you are left with a judgment that you must pay, the carrier might be required to pay the excess judgment too. The critical question to be answered is, "At the time and under the circumstances, was the policy limit demand reasonable so that it should have been accepted?"

In some instances, you, as the insured, may be required to contribute to the settlement of the case. This is a difficult situation for both the carrier and the insured, because the point of insurance is to protect the insured from paying for damages. If the case is a close call as far as coverage is concerned, or if there is a mixed bag of covered and non-covered claims, then contribution by the insured makes sense, provided the settlement resolves the entire case and there is no continuing liability. If it will not settle the whole case, don't do it.

If your carrier accepts the defense, it must defend all claims, covered and uncovered. However, if there is a reservation of rights at the end of the litigation, the carrier can file a *Buss* motion to recover fees and costs that it expended in defense of the uncovered claims. *Buss* motions are not typical, but they are an option for the carrier. This is why you want to make sure that if you contribute to the settlement, the carrier relinquishes all further claims against you; deductibles, SIR's, reimbursement under *Buss*, all claims.

When a Claim Is Covered

Determining when coverage is triggered can be a problem. For example, do you need a lawsuit to make it a claim? If the Owner contacts you and tells you the windows are leaking and are causing damage, but there is no lawsuit filed yet, is this a claim? It may be, or it may not be, depending on the way the policy is written. If the leak is causing damage, even without the lawsuit, you should notify your carrier and your agent that there is a potential claim.

Similarly, if there are other contractors whose work is implicated, you need to notify them. This is especially true in residential construction defect issues. Under the Right to Repair rules, subcontractors should be notified of any homeowner claims. Whether the problem is on a residential or any other type of project, it is always advisable to notify everyone with potential liability exposure. Always put your carrier on notice. Your policy will likely require that you inform the carrier whenever there is a claim or a potential for a claim. If you don't let the carrier know, you may be compromising your coverage. At the very least, talk to your agent and let them know of any possible claim.

Letters to the Insurance Commissioner

If you get the run-around from your carrier, or they are being uncooperative, or, more likely, non-responsive, then you and your attorney should consider a letter to the Insurance Commissioner. Insurance companies have statutory guidelines that they must follow when settling or dealing with claims. The guidelines, in broad-brush form, require fair claims handling procedures; i.e. timely communication to the insured of decisions, the bases for any coverage decisions, and reasonable and timely investigation of the claim.

If the carrier fails to comply with these basic obligations, the Insurance Commissioner can investigate and sanction the offending carrier. A complaint letter to the Insurance Commissioner is not your first resort but it is one that should be considered if you are not being treated fairly. Most carriers are quite good and will do what they need to do, but if they are not, or if the claims adjuster is, for lack of a better word, a jerk, then the letter to the Commissioner is probably a good way to get the carrier's attention. A letter to the carrier's home office complaining about the adjuster is also an option to get the attention of the people in charge. Any letter to the Insurance Commissioner or the home office should be well reasoned, thoughtful, well written, and well documented. Don't just vent; vent with articulation and eloquence.

Alternative Dispute Resolution (ADR)

If litigation commences, and often even before it commences, the parties will discuss ways to avoid the expense of trial. There are a number of ADR methods, but each has its own limitations and all are going to cost money, perhaps not as much as trial, but they will cost something. Therefore, if alternative dispute resolution is worth considering, it's worth considering early.

In certain circumstances, the option to engage in an ADR process is available only after the lawsuit is filed. For example in the public works context, if the prime contract claim resolution procedure does not specifically call for a particular ADR process, but instead proceeds according to the requirement of the Public Contract Code, then, depending on the amount in dispute, you will either be sent to mediation after you file your lawsuit or you will file your lawsuit and be off to trial. Either way, you need to file a lawsuit first. If it is a dispute with the State, you don't file a lawsuit; you go to arbitration.

In private works, the situation is a bit different. The parties are free to choose the way they want to resolve their disputes. It can be set out in the contract, or it can be agreed to after the contract is signed and a dispute arises. If the process is included in the agreement, it needs to be followed.

There are a number of forms of ADR: arbitration, mediation, neutral case evaluation, and combination procedures that mix litigation, mediation, and arbitration methods to get to an end result. You have the full selection of choices. Depending on the nature of the dispute, and the animosity of the parties, some are better choices than others.

Mediation

Mediation is non-binding; that is, no one can make you settle anything if you do not agree to the terms. There is no such thing as binding mediation.

With mediation, the first thing you need to decide is who the mediator will be, and this is an important decision. The mediator is a facilitator who will try to resolve the dispute between the parties. You want someone who has experience in the construction field, or at least knows something about construction and how things are done in the industry. Experience in construction law is also required, so that the mediator understands what the rules are and how they apply in the context of the case.

Mediation is the last best opportunity to resolve your dispute before it gets turned over to a judge and/or a jury (or an arbitrator). When it is in front of the judge or a jury, all bets are off. There is no guarantee the judge, let alone the jury, will have a clue about what you are talking about, which means your attorney will be both an advocate and an educator at the trial. As an advocate or an educator, the attorney must be persuasive, which means they need to understand the issues and be able to articulate them to the jury. This is not an easy process.

At mediation, you, the contractor, can articulate your position for yourself. Likely you will do it a lot more coherently than an attorney at trial. If the mediator understands construction, you will be presenting your case to a knowledgeable audience. Trials are not cheap, and they consume enormous amounts of time before the jury and just getting the case prepared, so if you can resolve the case with the help of a good mediator, it's money and time well spent.

A good mediator will also tell you what your case's strengths and weaknesses are so that you will get an objective evaluation of the issues in dispute. Sometimes, many times, after mediation the parties re-think their case completely and come away with a much better understanding of the issues, both theirs and their opponents. It is a great educational experience if nothing else, and is well worth the effort and cost.

The mediation process is completely confidential, and nothing that is said in the mediation can be used against the other party. The mediator will never be a witness at trial, and any settlement offers that are made will never be introduced as evidence. It is a completely confidential procedure.

However, just because you can't use the other side's statements from the mediation at trial does not mean that if you do not settle the case and your attorney takes the other side's deposition, he cannot ask questions that get that same response, and those can be used at trial. Therefore, be careful what you say and what you bring to the mediation, especially if it is shared with the other side. You do not want to disclose information or documents that might otherwise be privileged. The documents may remain confidential, but the information contained in those documents might be disclosed through deposition testimony. If the other side doesn't know certain information, there is no sense telling them that it exists and then letting them find a way to get it from you.

Remember, there is nothing binding about mediation. If you get there and the other side is being completely unreasonable, you can walk out. There are no consequences if you do, but it should be a good faith decision that drives your action.

Conversely, if you show up just to cost the other side money and to waste their time, then there may be consequences, especially if the mediator is reporting back to the court. The mediator can't reveal any confidential information that is shared, but the fact that one side or the other was wasting everyone's time and was not there in good faith can be disclosed, and the court may sanction the offending party. Sanctions are not good things to get. Therefore, if you don't want to do mediation, don't; but don't show up and waste everyone's time and money. Be there only if you want to be, and use the time to your best advantage by gathering information, and better yet, resolving the dispute.

Arbitration

Arbitration comes in two forms; binding and non-binding. The latter is, for the most part, not especially productive, and the former can be quite devastating. Arbitration is great for a lot of cases, but beware, there is no right of appeal in binding arbitration, so what you get is what you get.

> NOTE: If there is no arbitration provision in your contract, or if you are not required by Code to arbitrate a particular dispute, you cannot be forced to do so. Arbitration is either voluntary or it is contractual, but you cannot be forced to arbitrate a dispute. Likewise if you begin a litigation process and then want to move to an arbitration process, you may have a problem, depending on what has been done in the litigation process. A court may find that you have waived the right to arbitration, even if it is required by contract, depending on what you did before you demanded arbitration.

In arbitration, you and the other parties agree to an individual or a panel, usually 3 persons, who will hear the case and render a decision. The arbitration is typically held in a conference room, but the proceeding has most of the same formalities and attributes as a trial. The arbitrator(s) can decide how things will proceed, and the arbitrator(s) can make decisions about what is admissible and what is not admissible as evidence.

The rules of evidence are more relaxed, which is both good and bad. If you chose a particular arbitration service, American Arbitration Association, JAMS, ADR, or some other organization, they have their own rules for evidence and discovery.

If you sign up with one of these organizations, you agree to their rules, which may not always be conducive to the best presentation of your case. For example, depending on the value of your case, the discovery procedures can be quite limited. There may only be one deposition allowed without special permission from the arbitrator, and there may be limited document exchange. You will need to evaluate the evidence that you have against the information that you will need from the other side in order to prove your case. If you need a lot of discovery but the value of the case is low, certain arbitration services may not be the best choice. The rules can be modified by the parties, but make sure you do this before you agree to use any arbitration service, and make sure that the arbitration service will allow the changed procedure.

Binding arbitration is a bit of a roll of the dice. Even if you have a great case with all of the facts in your favor, you can still lose. It is the arbitrator's decision, and there is only a very limited basis for appeal. Unless the award is completely irrational and totally disregards the law and the evidence, the decision will stand, even if it is illogical and ignores key facts. You are done once the decision is rendered, so be sure of your case and be sure you know something about the arbitrator before you select him/her to decide your case.

Arbitration can only be required of the parties to the agreement; no one else is required to arbitrate. Others can agree to join, but if they have no contractual obligation to do so, they cannot be compelled to participate. The decision that is rendered is binding only on the parties to the arbitration, and the conclusions of law that are given are binding only on the parties to the arbitration. Sometimes, but not often, non-parties can use those rulings against you in another proceeding, but unless those other parties were participants at the arbitration, and not just witnesses, you cannot use the rulings in the arbitration against third parties that did not participate.

The point is that if you are intending to arbitrate a dispute, make sure that you have all the parties at the arbitration that are necessary for the complete resolution of the dispute. This includes all sureties and insurance companies, if they are involved and/or have potential liability. This is especially true and critically important where, for example, there is a bond claim, but the surety is not a party to the arbitration. In this situation, if the arbitrator finds in favor of the other side but does not award attorney fees, your opponent can still sue the surety for those fees if the surety did not participate.

This completely circumvents the decision of the arbitrator, because the surety has an indemnity claim back to the principal, and the principal will pay the attorney fees that the surety is required to pay, even though the arbitration award said no attorney fees were awarded. Be very careful when you arbitrate if there are bond claims involved. Be sure to have all parties present, including the surety, even if it is just by including the surety by name in the arbitration. Be doubly sure that any judgment and ruling includes, and is binding on, the surety as well as you, the contractor/principal.

Non-binding arbitration is a process that can be engaged, but the arbitrator's decision is not binding. If you don't like the decision, you ask the court for a trial de novo, which translates to "do-over." The process may not be binding, but it is definitely instructive. You get to hear the other side's case, and they get to hear yours, so that when you finally get into court, both sides are teed up for the long shot, having already played the course once. If you have the money, the inclination, and the time, by all means have a non-binding (judicial) arbitration. If you have other things to do than being in court with your attorney, don't waste your money with this process. Sometimes, very few, it works to resolve the issues, but for the money, mediation is a much better option.

Mixed ADR Services

Mixed ADR processes are also available. There are some organizations, for example *ArbitrationWorks* and JAMS, that combine some of the better features of arbitration and mediation with the more practical aspects of construction disputes, like schedule analysis and document review, and put them into one great glob of a process that gives out a decision at the end. The parties must agree at the outset whether the decision is binding, and whether there can be an appeal. If the process is not binding, don't waste your time, but if you use it, make sure you include the option of appeal.

In this process both sides give all the documents that support their claim, and all relevant project documents, to the third party facilitator (the "TPF"). The TPF usually has a Code specialist, a scheduling specialist, an attorney, a contractor, an architect, and an engineer working with them; anyone and everyone who would be needed to review the documents and understand the issues. These people go through the documents, re-create the schedule, analyze the delay, re-create the critical path, find the impacts, analyze the costs of the impacts and delay and inefficiencies, analyze the costs of disputed issues, and then render a decision with a value assigned as to who gets what for what.

It is a time-consuming process, and relies on the cooperation of the parties to make it work. Cooperation, of course, is not always possible. More important, one or the other side may have the better records, and as a consequence will sometimes get the better result. Why? Because, the more documents there are, the more complete will be the support of that position. If you have the better documentation, your position will appear more logical, and presumably the determination will be in your favor.

These types of mixed ADR evaluations can be quite pricey. They are most productive on large construction claims, and they are often employed most effectively during the course of the project as opposed to at the end, after the sides are polarized and lines are drawn in the sand. This is a good process where rational and reasonable people are present on both sides.

Settlement

Settlement of a dispute is always good, but once the settlement is reached in principal, it needs to be, and should be, confirmed in writing. The devil is in the details, so an attorney should prepare any settlement agreement. The settlement agreement is a contract, and it has the full force and effect of any contract. It needs to be prepared correctly, it needs to give you all the protection you are entitled to, it should not include any terms that were not agreed to, and you need to understand the consequences of the settlement agreement, just as you would need to understand any other contract.

Are You Out of the Case?

The first question to ask with any settlement is: will settling now take me and/or my company completely out of the case? Sometimes the answer is NO, in which case it may not be in your best interests to settle.

For example, assume you are a subcontractor and it is a construction defect case. Your work is alleged to have caused damage, and you want to pay some money and get out of the case. Unfortunately, your contract with the General has an express indemnity provision in it.

In this instance, even if you settle you may still have obligations to others for indemnity that are not released by the settlement, even if you pay every dime the other side wants. Therefore the first thing you want to confirm is, "Am I completely out of this litigation, or is there continuing liability that I still need to deal with?"

Change the hypothetical and assume you have no express indemnity obligations to anyone, and you want to settle with the other side. You are one of many defendants, and all defendants have sued all other defendants for implied indemnity. Can you get out? Yes, but you need to bring a Motion for a Determination of Good Faith; that is, you want the court to review the settlement and conclude that it is fair and equitable and that you have paid a fair share of any potential liability. You want this determination from the court so that you are completely out of the case.

There are a number of factors that the court will review; the claims against you, the claims in general, the amount the Plaintiff is suing for, the amount you could be liable for, and the amount that you are paying to get out of the litigation. If the amount is disproportionately low, or if there is collusion going on, the court will deny the motion. This does not mean you can't still settle, but the claims of the other parties for implied indemnity do not get dismissed and you will still be in the case. If the motion is granted, you are out of the case.

Dismissals, Releases & 1542 Waivers

In any settlement, the main thing you want in return for your payment is a dismissal. You want the dismissal with prejudice, which means that this is the end of the lawsuit, and all the issues are resolved. You do not want a dismissal without prejudice! Be very clear in the Settlement Agreement that the dismissal is with prejudice. This may come as a great shock to you, but sometimes attorneys draft documents that say things that are different than what they told you they would say. Read the Settlement Agreement as you would any other contract.

The other critically important thing you want in a Settlement Agreement is a release from all liability for your insurer, surety, and any other agent or representative you have. Recall the instance we discussed in the arbitration section where a surety can be sued for fees after the arbitration concludes, and that those fees will end up being paid by you, the principal, even if the arbitrator did not make an award of fees. This is also the case in Settlement Agreements. If the party or its representative is not expressly released, the party or representative is potentially still liable. Make sure to include, by name and description, all the parties that could have liability, and that you might need to be responsible for, in any Agreement.

Settlement Agreements will also include what is referred to as a 1542 waiver. Civil Code section 1542 provides that only the claims that the parties know about are settled and released in a settlement, and any claims that are unknown are not being released. The 1542 waiver in the

Settlement Agreement states that the parties are waiving that provision and that they intend by their Agreement to give up all claims, known and unknown, except any that are expressly reserved by the terms of the Settlement Agreement.

In other words, when the Agreement is signed, the litigation and all issues related to the litigation are concluded. If you want to carve out specific issues that are not being released, you need to specifically include those exclusions in the Agreement. The typical issues that are reserved and not released are latent defects, warranties, and indemnity for 3d party claims. There may be others that are important to the parties, and these should be negotiated and discussed, and included in the separate provision for exclusions.

> **Note:** You do not want an issue to be reserved if there is nothing to be gained by your reserving it. Do not include issues that you don't want to deal with and are simply putting off. Be very clear in any carve out exactly what is being excluded. For example, do not exclude "the issues related to the HVAC." Itemize those issues and/or include a separate Exhibit that identifies exactly what is included and who has the responsibility for whatever the issues are. Do not leave unclear and ambiguous loose ends unless you want to have further disagreements later on. You want a limited number of issues excluded, because otherwise the Agreement is not really doing very much.

Recitals- Long or Short

Recitals are the introductory summary of facts that gave rise to the dispute in the first place. Recitals can either be quite extensive or very cursory. What they should not be is pejorative; that is, they should not be accusatory, but rather a simple statement of the issues or events.

Be careful with the drafting of the Recitals, because the wording may make sense in context at the time the Agreement is prepared, but later, when the Agreement is looked at from the distance of a couple of years, new interpretations or meanings can sometimes be read into the words that were used. Review the Recitals as if you were looking at them without any idea of what the dispute was about. Do they describe the circumstances? Are they clear? Can you twist the words around to mean something that you do not intend them to mean?

As a practical matter, the best Recitals are often the briefest and most concise. The Agreement is a contract to settle the case, not argue it. The terms of the settlement are what matter, and the underlying issues are not that important anymore. Obviously there was a dispute, and that is what you are settling, but the details of the dispute are really secondary to the terms of the Agreement that settles the dispute. The terms are what need to be clear in the Agreement.

Other Terms to Include In the Settlement Agreement

An important provision to include is the representation that all parties to the Agreement agree that they have each contributed to the preparation of the document and it is not just one party's work. If the terms of the Agreement are ever disputed, then no one party can be charged with creating the ambiguity. In California, ambiguity is construed against the person who creates it. Therefore, if everyone contributes to the document, no one side has responsibility for the ambiguity. Likewise, there cannot be any claim of undue influence. If you don't have a say in the language or don't agree to the terms, don't sign the agreement and don't include this type of provision.

Be sure to include where any dispute related to the Settlement Agreement can be resolved. Typically, the court that will have jurisdiction is the court in the county where the project is located or where the underlying lawsuit was filed. If one location is going to have exclusive jurisdiction, be sure to identify it. Make sure that California law applies to the terms of the agreement, especially if one of the parties is from out of state.

Include a provision that the persons signing the Agreement have the authority to sign on behalf of the party they represent. Make sure that there is a signature line for all parties to sign and date. You should also include a line for attorneys representing the parties to sign that states that they agree to the form of the Agreement. If no attorneys are involved, then include a line that states that all parties have had an opportunity to consult with counsel and understand the terms of the Agreement they are signing.

Finally, as with all legal documents that are multiple pages, initial all pages and keep a copy of the Agreement that has your initials. If there are attachments, make sure those attachments are identified and included by reference in the Settlement Agreement, and initial those attachments, too. The Agreement will be valid when all parties have signed and/or the required acts are completed; i.e. money is paid, dismissals are entered, etc.

> **NOTE:** It is good practice to include a requirement in the Agreement that you will receive a copy of the signed Dismissal with prejudice that will be filed with the court as a precondition of any payment you will make. You also want a copy of the endorsed court filed Dismissal (the copy that is filed in the court) for your records. Make sure you get it. If there are Stop Notices or Mechanic Liens that are on file or recorded, you will want copies of the Stop Notice Release and/or the Mechanic Lien Release, and the Release of any Lis Pendens. You will also want signed statutory Conditional and Unconditional Lien Releases Upon Final Payment.

CCP 998 Offers

A party can offer to settle the case with another party at any time and in any manner they want. An offer to compromise can be as simple as a letter to the other side, or it can be a formal statutory offer made pursuant to Code. All offers to compromise are privileged and cannot be introduced at trial regardless of how they are made.

Statutory offers to compromise, those made pursuant to Code of Civil Procedure section 998 (the '998 Offer'), are complicated. They can be a drafting nightmare, and they have certain serious repercussions that need to be considered before they are made. Either plaintiff or defendant can make them, and depending on who makes it, the consequences will vary.

If the 998 Offer is made and is accepted, the case is over. If the Offer is not accepted, it will terminate after 30 days and is deemed withdrawn. This is where the fun begins, because there are consequences for making the Offer and there are also lots of potential issues with the way the Offer is made.

If the Defendant makes the 998 Offer and it is not accepted by the Plaintiff and the Plaintiff does not recover more than the Offer, the Plaintiff cannot recover its post offer costs as damages and must pay the Defendant's costs from the time of the Offer. The costs allowed under the section, from the time of the Offer, are also deducted from the amount the Plaintiff recovers. Similarly, if the Defendant does not accept the Plaintiff's Offer and the Plaintiff obtains more than the Offer, the defendant may have to pay certain expert costs.

This all seems simple enough, but it isn't. For one thing, how the 998 Offer is worded may cause problems even if it is accepted. For example, if the Offer provides that the party making the Offer will pay a certain amount to the other side and will also give up all claims to costs and fees that might be recoverable, and the other side accepts it, that may not be all the accepting side can get. The Offer provided that the offering party would give up its entitlement to fees and costs, but it did not require that the other side would also give up costs and fees. Therefore, the court could determine that the accepting party is the prevailing party, and if there is an attorney fees clause in the contract or available by statute, the accepting party may be awarded its fees and costs in addition to the money paid to settle the case. Statutory Offers are complicated and tricky, and they need to be considered carefully and worded appropriately.

The 998 Offer must be unambiguous as to what it is offering, and there must be a way to value the offer if something other than money is being offered. If the Offer includes, for example, an assignment of rights for claims against another party, and if the Offer is not accepted, then when the trial is over the court must be able to put some value on the assignment of rights in order to determine whether one side or the other recovered more (or less) than the Offer. If there is no way to do that, the

998 Offer is unenforceable because it is too vague. The 998 Offer has to be susceptible to clear valuation, or least reasonably clear valuation, in order to be effective. If it is not, the Offer is not going to have any use and will be ignored.

When to Make a 998 Offer

If your case is strong and the facts are clear from the beginning, the Offer can be made early on. Typically the 998 is made before any expert depositions are taken, so that the cost of those depositions and any other related costs can be passed on to the other side if they reject the Offer and they do not recover more than the Offer. Experts are expensive, and are usually deposed toward the end of the litigation, usually beginning 60 days before the trial date. This is an added incentive to accept the Offer, because both sides avoid the need to incur expert costs.

As a tactical matter, the Offer should be an amount that the offering party calculates the other side would be willing to accept to avoid the cost and risk of trial. These costs are measured against the amount the other side is seeking and the likelihood that they would recover that amount at trial. Everyone has their own way to evaluate risk, and discounting that risk for uncertainty, and this analysis will be used to structure the 998 and to evaluate any that you receive.

Low-ball offers and very high offers are pointless and counterproductive. They set the bar too low or too high, and if the jury awards a higher amount, the offer gives the other side a significant benefit and advantage; they may get certain expert costs. Statutory offers are intended to encourage settlement, and they do this by making the consequence of undervaluing or overvaluing the case fairly significant.

Acceptance, Multiple Offers, Partial Payments

A 998 Offer must be accepted within 30 days of when it is made or it is deemed withdrawn. The Offer can be made up to 10 days before trial, but it must be accepted before trial commences. Multiple 998 Offers can be made, and often are, as the dynamic of the litigation changes. Each new 998 is the operative 998 and all prior Offers are deemed withdrawn (see Note 25 for a new case that may change this). A 998 can be withdrawn after it is made if the other side has not already accepted it. Finally, if there is a partial payment of the amount in dispute, the operative 998 Offer is invalid, but again, another 998 Offer can be made. The statutory Offer is a serious matter though and multiple 998 Offers are not especially useful.

Terms Used in Litigation

If you are involved in litigation, there are certain terms that you may hear bandied about by your attorney or some of the other parties. The following is a brief description of what those terms mean and how and when they might come up in the litigation.

Demurrer When a Complaint is filed, it must state a cause of action. In other words, the elements that make up the claim must be present in the Complaint. If the Complaint contains a lot of facts, but all those facts, taken together, do not make up a cause of action, you will demur.

Essentially, the demurrer says that even if everything you say is true, you cannot state a cause of action against me. If you demur to a Complaint, and in some instances to an Answer, you are basically pointing out the flaws in the other side's argument. Unless there is absolutely no way for the other side to amend the pleading to add new facts to the Complaint so that it will state a viable claim, it is better not to waste the time and fees that it will take to demur.

A court is likely to agree that the Complaint lacks the proper facts, but the court is just as likely to give the other side leave to amend the pleading to see if they can cure that defect. It is only in the case when it is obvious that the flaw in the Complaint cannot be cured by amendment that the court will not grant leave to amend; i.e. the statute of limitations has run out, or the Complaint admits facts that undermine all the allegations that are included.

Motion to Strike When the Complaint makes an allegation that is intended solely to embarrass the other party, or is so outrageous that it is insulting and despicable, or the allegation is not necessary for the cause of action to be stated, the court may order the Plaintiff to strike that allegation and amend the Complaint so that it is not included.

Likewise, if there is a reason why the Complaint or an Answer should not be filed in the first place, then a motion to strike is permitted. For example, if the Plaintiff or answering defendant is a suspended corporation, it cannot file a lawsuit or answer a Complaint; it cannot participate at all, so a motion to strike is appropriate.

Default After you are properly served with a lawsuit, you have a specified period of time within which to answer. Most often it is 30 days, with 5 added days if you are served by mail. If you decide to ignore the Complaint and don't do anything with it, and if you ignore letters from the attorneys that say that you need to Answer, then a default may be entered against you or your company. A default basically says that you have conceded the dispute and admit you are liable. The Clerk enters the default, the other side then asks for a judgment, there is a prove-up hearing where the other side will establish the amount you should pay, and the judge can then enter a judgment against you.

It rarely gets that far. If a default is entered, you need to file a motion to set aside the default. The court will want an explanation about why you did not answer. It needs to be a reasonable one. Similarly, it is not advisable to serve the Complaint, wait 30 or 35 days, and then run to court and take a default. The courts do not like that type of un-professionalism and gamesmanship. However, this is not to say that the court will always set aside a default. If the facts are such that a default is appropriate, the court will let it stand and the defaulted party would need to file an appeal, which is costly and time-consuming.

Summary Judgment-Summary Adjudication During the course of the litigation, the parties will conduct discovery (ask for documents, send interrogatories, send requests for admissions) so that they can be ready for trial. Sometimes it becomes apparent that one side either has no case to assert or has no defense to the case against it. If there is still time, because these types of motions must be filed 105 days before trial, the party that thinks it can win on the basis of facts and admissions that exist and/or are revealed during the litigation can move for either summary judgment or summary adjudication.

Summary judgment will conclude the entire case. If you win on summary judgment, the case is over. In order to win summary judgment, there can be no dispute over any of the material facts that serve as the bases for the causes of action in the Complaint. In other words, every material fact, a fact of consequence in the litigation or to the cause of action, must fall in your favor.

These are complicated and very time-consuming motions that are "disfavored" by the courts, because the point of litigation is to get the case to the jury so that all the issues can be decided before an impartial panel. If you can show that there is no need for a trial because all the undisputed material facts demonstrate that you win, you should get summary judgment. It's a lot cheaper to win this way than to go to trial and present the evidence to the jury.

However, in order to defeat summary judgment, all you need is one disputed material fact. One! This would seem to make summary judgment hard to obtain, and it is, but under the right set of facts it is well worth the effort. Judges are, for the most part, pretty smart folks, and they can see through a lot of BS and smoke if they need to in order to reach a finding that summary judgment should be granted. The court must weigh all the arguments and facts against the summary judgment with great liberality, and all the arguments in favor of it very strictly. It is not an easy motion to win, but if you have the facts it is worth the effort.

If your attorney suggests summary judgment, ask lots of questions, because these are very costly motions. If there is any questionable or disputed fact, don't move for summary judgment. Do not let your attorney convince you with a bunch of pie in the sky wishful thinking. Make sure it's not just a good argument, but an air-tight one, and get a budget for the work. When you get the budget, you should probably be sitting down when you read it.

Summary adjudication is slightly different and a little easier to win if you have the facts. Summary adjudication is not an all or nothing motion, but rather it attacks each of the causes of action and tries to knock them out one by one by showing that there are no disputed material facts for each claim, and therefore you should win as a matter of law. If the Complaint has five causes of action and you get summary adjudication on four, you are still in the litigation and headed for trial. It may be an easier and faster trial, but you are not done with the case.

The other side beats summary adjudication by showing one disputed material fact for each of the causes of action, just like with summary judgment. Again, the analysis starts with the question to your attorney, "What will be accomplished, assuming we win on any of the causes of action?" If there are a number of causes of action in the Complaint, but the motion will only knock out one of the less important claims, the question is, "Is it worth it?" Probably not. You are still going to trial, and the other side still has a case. Summary adjudication is an expensive motion, just like summary judgment, so save your money for trial.

Typically, motions for summary judgment and summary adjudication are filed together, and as alternative requests; give me summary judgment or, in the alternative, summary adjudication for one or all of the causes of action. The cost of doing both motions is not quite double, but the preparation and drafting requirements will consume a lot of attorney time. If you have a strong argument against only one cause of action, then move for summary adjudication against that one claim only, especially if the one claim is a potentially expensive proof or liability issue.

For example, if there is a claim for punitive damages but the facts of the case don't support it, then moving for summary adjudication is a good idea. It takes away a potentially very costly claim, it reduces the elements of proof, and therefore the time at trial, and it may take some of the wind out of the other side's sails if they lose.

Bottom line; there will be a time before the trial when the court will want the parties to whittle down the issues, so consider your options wisely. Do you really need to spend the money now to knock out a claim that is not a really important issue at all? Ask your attorney hard questions about the likelihood of success, what is accomplished by a win for the individual cause of action, how much it will cost, and what it will save at trial. These are great motions to make if you can win them, but they can be a colossal waste of time and money if you can't.

Discovery This is where you get to find out what cards the other side has, even if you will not necessarily know how they will play them. Discovery is a pretty expansive right in any Unlimited jurisdiction case, is fairly constrained in Limited jurisdiction cases (those $25,000 or less in value), and is non-existent in small claims court actions. The methods of discovery available are depositions (testimony under oath by questions and answers), special interrogatories (written questions sent to the other side with answers given under penalty of perjury), request for admissions (questions sent that ask you or the other side to admit the truth or falsity of a particular fact or document), and inspection demands for either documents or property.

Depositions and document inspections are usually the most productive means of discovery in construction dispute litigation. Site inspections are also usually necessary and helpful. Special interrogatories are OK if you need some names or want to know a particular fact, but otherwise the answers are orchestrated, if not totally prepared, by the other side's attorney. Not especially helpful. Requests for Admissions are good if you have a lot of documents that you want to have authenticated before you take a deposition, and if you have some important questions about the truth or falsity of certain propositions that may be at issue in the litigation.

Discovery takes time to prepare and answer, and it can become very expensive indeed, so meet with your attorney and discuss what the important facts are that need to be developed, so that the attorney is not just out on a fishing expedition. A document review of the other side's files is a time-consuming and tedious effort that needs to be done with a plan and an end purpose in mind. The attorney should be the one who reviews the documents of the other side, but he/she will need to discuss with you what documents to look for. The best document review is often conducted by the attorney with you (your project manager or superintendent) assisting.

Document review should be of all the files the other side has, and you should decide what you want to see. Don't let the other side decide what they will show you. Unless there are only a few specific documents that you can identify by name, you should not rely on the other side to give you all that you request. You want to see the job files, the plans and specs, the contracts, and any daily logs, notes, journals or other project-related writings. These can be gold mines of information.

Likewise, depositions are important. If you can attend to assist the attorney, that might be helpful, but you should ask if it is necessary. Often witnesses will not be as free and unguarded if they see the other party across the table from them. In the deposition, too, attorneys often point and shoot and hope they either get something or stir up something that gets them something. They may not have a real plan, and your presence may distract the attorney. Depositions should be planned ahead of time, but sometimes the most important information obtained comes from the most innocuous questions. It is usually better if the client is not at the deposition, but that is not always the case. If nothing else, the client should be involved in helping the attorney prepare for the deposition.

Discovery is another area where you need to contain costs. Get a budget, or at least an outline, of the type of discovery your attorney is intending to undertake. Work with the attorney to put together a list of necessary depositions, what documents are important to obtain from the other side, and what the basic facts of the dispute are, so that your attorney knows what he is going after when he conducts discovery.

Avoid the client-sponsored fishing expedition. Get a plan together, explain to your attorney what went on during the project, and make sure the attorney explains to you what claims are being made against you, or that you will need to prove against the other side, depending on your role as plaintiff or defendant. Your plan needs to be flexible, because it is often the case that discovery reveals documents or testimony that will completely change your approach to the litigation, or will modify your strategy in a very profound way.

Likewise, if the case is something that your insurance carrier has picked up, you need to cooperate with the appointed attorney. Most of what was discussed above will apply. Your case and the litigation costs are something that you should be concerned with, and assisting the attorney in discovery and with other issues is something that is important, regardless of who is paying.

The Litigation Privilege

The litigation privilege attaches to protect the parties in the litigation from any liability for things they say in the litigation that might otherwise be actionable. If you are sued, you can pretty much say anything you want about the other side, provided it is at least somehow related to the case at hand, without being concerned about getting sued for slander or libel. However, if there is no litigation and you start slinging mud, there could be liability.

This is important to remember in public works contracting. Recall that the False Claim Act ("FCA") applies to claims that are made to the Owner before there is litigation, and that are intended to induce or are capable of inducing the Owner to pay money that it is not otherwise required to pay. If demands are made to the public agency before litigation, and if they are false, then the FCA can apply and it can be nasty.

On the other hand, if you sue the entity for a certain amount, even if the amount is exaggerated, there is no false claim and the claim is protected by the litigation privilege. In those circumstances, the pre-litigation claim and the litigation claim amount should be reasonably close, if not exactly the same. If you admit at trial that the pre-litigation claim was exaggerated and knowingly false, then the public agency can amend its complaint or cross complaint to allege an FCA violation.

Important Files to Keep

When you close out a project, Business & Professions Code section 7111 requires that you keep your job records for 5 years. It's a better practice is to keep your job records for 10 years, so that in the event you get sued for latent construction defects during that time, you will have the documents you need to defend yourself. Certain documents are more important to have handy than others. You want/need to have these critical documents assembled and archived so that you can easily access them. The following is a list of some essential documents that you should keep, especially if you archive your files in a warehouse or if you purge your files after 5 years.

For the General Contractor:

- The Prime Contract, including any escrowed bid documents.
- All subcontracts, and a copy of the boilerplate subcontract.
- All proposals from any subcontractors or suppliers.
- Your bid file from bid day for the Project.
- All subcontractor Insurance Certificates and Additional Insured Endorsements. *THIS IS CRITICAL.*
- The Notice of Completion or Cessation, the Notice to Proceed.
- RFI Logs and any other Log that will describe changes made during the course of construction, as well as the as-planned and the as-built schedule, punch lists, change order logs, ASI logs.
- The project plans and specifications, especially any annotated plan set or the job copy. Bid set, job set, as-built set.
- Photographs of the Project, especially when the Project is completed; interior and exterior finishes.

For Subcontractors:

- Your subcontract, the specification section you bid, any change orders, all revised drawings.
- Any proposal that you provided to the GC.
- Plus all of the documents listed in the GC section that you may also have.

If litigation has followed the completion of the project then you should have a file or binder that contains:

- The Complaint, Answer, and any Cross Complaint.
- Settlement Agreement; fully executed (signed by all parties).
- Dismissal; endorsed copy from court.
- Releases of Liens and Stop Notices, if any.

SB 800 & Calderon

In response to the flood of construction-related litigation that occurred during the 1980's and 1990's and into the early 2000's, the Legislature undertook to make some comprehensive and fundamental changes to the law related to residential construction, specifically production or tract housing. The developers of these types of projects, the insurance carriers, and the attorneys for the developers and for the plaintiff side homeowners got together with the Legislature and came up with a new way to do business. The result was the Right to Repair Act, also known as SB 800, which is found starting at Civil Code section 896. If you do not work on residential single family home projects, this Code section should not affect you. If you do, then it will, and it can have some significant implications.

SB 800 was intended to cut down on the amount of litigation that followed the completion of most, if not all, large residential tract housing. The lawyers that represented homeowners and filed the lawsuits were pretty industrious and creative in their approach to this litigation. In some instances, they would actually put fliers in the mailboxes of the homeowners advising them that one of their neighbors was experiencing problems with their home (leaks, cracking, settling) and that the law firm was being retained to represent those individuals. The neighbors were advised that their homes might be experiencing the same problems, and that these attorneys were contemplating a lawsuit against the developer so that they could get these problems fixed. If the neighbors cared to hear more, they could come to a meeting at a particular time and place and hear what the attorneys intended to do and how they intended to do it.

What followed, of course, was a lawsuit where the Homeowners Association or a group of individual homeowners sued the developer. The plaintiff's attorneys would have "consultants" review the conditions and undertake destructive testing on the various homes in the subdivision. Invariably, they found problems, and just as invariably, the cost of the repair of those problems was an astronomical amount.

When the HOA filed its action, the developer would sue the General and/or all the subcontractors involved in the project. The insurance carriers for all the contractors were put on notice, and the defense attorneys were rallied and assigned the case. The lawsuit was off and running.

In short, it was unfair, and construction defect litigation did not do much except make attorneys and "experts" a lot of money. SB 800 was intended to fix that problem, but whether it did or not is still an open question. Contractors now have a right to repair the work, but even if they do, it may not be the end of the issue; there is no release of liability for only repairing the alleged deficient work, there must also be a payment of money to the homeowner.

The Right to Repair Act provided a series of "functionality standards" that residential construction work is expected to meet. These standards are laid out in the language of the statute itself, and if you are doing work on multi-unit residential, which is the only type of construction this statute applies to, then you need to review the standards and the impact they will have on your trade.

The standards are fairly vague, and include terms and descriptions like "excessive condensation," or "materially impairing" the use, or requiring that systems must "operate properly" and not create "unreasonable risk." It's hard to know exactly what these words mean, but those are the "standards." Not all the standards are so poorly stated, but if you work on these types of projects, you want to know what the standards are for your trade, however poorly worded that standard may be.

NOTE: SB 800 created a new form of strict liability where a contractor can be responsible for a breach of the functionality standard, regardless of whether the failure to meet the standard causes damage or not. Depending on the trade, these standards can be problematic, not just because of the vague definitions that are included, but because with no damage required to trigger a breach, the issue of insurance coverage for these claims becomes problematic. Recall, insurance is only triggered when your work causes damage to other parts of the project. No damage could mean no coverage, and the contractor might be left to fend for itself.

SB 800 requires that the homeowner provide the General/developer with notice of the problem, and the General/developer has a prescribed timeframe to investigate and propose a repair. These initial timeframes are ridiculously short, but they can be extended. If you are a subcontractor, it is unlikely that you will get the original notice, but instead will be informed of the problem and any subsequent site visit, so that you can review the alleged defective condition. This may be a problem if the damage is ongoing or could have been prevented earlier, or if the situation changes by the time you see it.

Once the contractors review the situation, and assuming they chose to make the repair, that will still not be the end of it. You can still be sued, and the fact that you made the repair at all can be introduced at trial to show you did poor work in the first place. If the repair you make does not fix the problem, you are doubly damned.

If you want to settle the case once and for all, you need to make a payment that will either cover the cost of repair and any other issues the homeowner has, or you can make the repair and also pay money to settle the case. Unless you pay, though, the homeowner does not need to give you a release and it is still possible you will end up in court.

SB 800 also changed the statute of limitations for certain kinds of claims. In some cases this is good; in others not so good. Some of the new statute of limitations are fairly short, and if the homeowner does not give you notice in that time frame, they cannot raise the issue at all.

You should review the time limits in the Code that applies to your trade, although as a practical matter, you will not need to worry about these limitations; your attorney will. Keep good records of when you worked on the project, that is, which house in the project on what day, and, if you had any issues, what they were and when you let the General/developer know about them.

Some final points related to SB 800.

First, it does not really matter whether the Owner gives notice or not. If the homeowners file suit prematurely, the court can stay the proceeding and allow you to inspect and repair. Of course, it really makes no difference whether you do the repair or not, the lawsuit can continue in any case. Only if you pay money are you out of the case.

Second, the requirements of SB 800 only apply to residential construction, and to those residents with a close of escrow that occurred after January 1, 2003. If the close of Escrow was before that date, SB 800 is not relevant.

Calderon. If the project you worked on was completed before January 1, 2003 a similar type process, the Calderon procedure, is implemented. In this situation, the homeowners and the contractors are required to meet and confer and share information before litigation is started. The Homeowners Association is required to provide Calderon notice to the developer or General Contractor, which triggers a series of required events and meetings.

If the HOA fails to provide the notice but sues instead, the court will fashion the necessary relief, typically by directing the meeting/investigation process to commence. The parties will need to share information, and the proceedings will be stayed until these preliminaries are completed. Alternatively, the parties can stipulate to forego the Calderon process or they can engage in some abbreviated or alternative proceeding.

Under either SB 800 or Calderon, you need to tender the claim to your carrier, just as you would any other claim. You do yourself and your attorney a great favor if you have kept good and complete records of what went on and what you did. Pictures are a great way to preserve the moment, but if you take pictures of your work, make sure that the photos are consistent with the plans. No sense having your photos show that you did not follow the plans!

SB 800 and Calderon did not do away with the Plaintiff attorneys' interest in this type of litigation. It remains as lucrative to them as it always was. If you are building these types of projects, do them right and make sure you have good records.

Chapter 9

Employment Issues Briefly Noted

IT IS NOT DIFFICULT to obtain a contractor's license, and there are a lot of people who have them. However, just because someone has a license does not make that person an independent contractor that you can hire, pay directly, and 1099 at the end of the year. An independent contractor is a very carefully defined term, and unless most, if not all, of the factors are met, the person you hire as an independent contractor may not be independent at all; they may be your employee and entitled to back wages, overtime, SSI, and other required contributions. You may also be responsible for significant penalties from the IRS, among other agencies, if you mischaracterize an employee as an independent contractor. Keep in mind, too, that the employee relationship is the favored finding, so unless the independent contractor is truly independent, a court is going to find you have an employee, not a contractor.

The Independent Contractor

First and foremost, the independent contractor must be independent. The first factor, then, is who has control over the means and methods of the performance of the work. It does not matter if the person you hire is told that they can do the work however they want. If they cannot really exercise that right, there is no right, and they are your employee.

The next consideration is whether the worker is engaged in a distinct occupation or has an independently established business. In other words, is the independent contractor really providing a service as their business, or are they just providing it to you? For example, if you hire someone as your superintendent on a project and they have a contractor's license, are they an independent contractor? Probably not.

For one thing, the person cannot superintend more than just your job, so it is not likely they are an independent business. They will also be constrained by the project; when it is over, they will be out of work. There is nothing really independent about the contractor. Conversely, if you contracted with a company that provided superintendents to projects, then the person on your project would not be your employee, although they would be an employee of the other company and the other company is an "independent" contractor.

Another criterion is whether the independent contractor or the principal supplies the tools or instrumentalities used in the independent contractor's work. The independent contractor has to bring something to the job, not just him or herself. If you are providing the trailer, phones, faxes, desks, power, etc., and the person just shows up and uses those facilities, then likely the person is not an independent contractor and is an employee.

How do you pay this person? Are they paid weekly, or is there some other arrangement? Is it by the job or is it by the hour? Does the person have the opportunity to make other money as part of this business, and do they? Does this independent contractor have employees, or is he/she a one-person shop? Can they set their own hours, go on vacations when they like, and perform the work however they want? Do they have a separate business address and license with the local business bureau? How permanent is their relationship with you?

All these considerations are relevant to the determination of whether you have an independent contractor or an employee. In short, if it looks like an employee-employer relationship, it is. There have been court cases where the relationship appeared to be very clearly an independent contractor relationship, but the courts found to the contrary. For example, in one instance the court found that delivery persons with their own trucks who could set their own hours and routes, were actually employees of the company. The company was then hit with overtime claims, penalties, and taxes.

The point is that if it is a close call, you should opt for the employee-employer relationship and treat it as such. The money saved in the short run is not worth it if the government catches up with you, and, even more aggravating, if the so-called "independent contractor" gets a lawyer and pursues you for overtime and other claims because the so-called independent contractor insists you were their employer and that they are owed for overtime, social security, vacation, retirement, and any other benefit that an employee would ordinarily receive. It's a scam that is being used, and used successfully, so be careful in any independent contractor situation.

If you do decide to hire an independent contractor, then do not perform yearly evaluations, request notice of vacation days, keep track of sick days, provide an employee manual, or insist that the independent contractor must conduct themselves in a way that will reflect well on your company because they are in some manner a representative of your company. All of these sorts of things are employee-related activities. If you have an independent contractor relationship, you must maintain it, and not compromise it by reverting to the employer-employee relationship.

Better still, just hire the guy.

Chapter 10

OSHA: When Bad Things Happen To Good People

LET'S FACE IT; CONSTRUCTION is an inherently dangerous profession. Not only is the actual construction site filled with any number of opportunities to cause physical injury or worse, but the tools that construction workers use can do some really nasty damage, even if they are used in an appropriate and totally correct way. If the job site or the tools, or both, are not maintained in an appropriate manner, then bad things happen, and when bad things happen on a job site, OSHA will become involved. OSHA can become involved even if bad things have not happened.

When OSHA Comes Calling

What happens when OSHA gets involved or just shows up for an inspection? What do you need to do, and what does the inspector have a right to do? How do you deal with the inspection or investigation?

First, you cooperate. Whether it's an investigation because there has been some job-related injury or just a periodic inspection, you need to work with the inspector. It does no good to stonewall the inspection, deny access, or become confrontational, especially if something really bad has happened on the site. You will have the opportunity to contest any adverse findings later, but if the inspector has presented his credentials correctly, the best approach is to work together, not in opposition. Good common sense and discretion are the order of the day.

Required Notice of Injury

Whenever there is a death or serious injury, OSHA must be notified immediately. Your field personnel need to know what to do in this instance, and how to get in touch with OSHA. If emergency responders are called to the site, OSHA must also be notified regardless of the degree of injury. The responders have an obligation to notify OSHA, so one way or the other, OSHA is notified. The fact that the EMT informs OSHA does not relieve you of your obligation to notify. If you don't call, that is a violation, and you will be cited for it.

If your main office gets the call from the field that there has been a serious accident on the site, the office must also notify OSHA within 8 hours. If for some reason your office does not know of the injury immediately, you will have slightly more time, but once you know that there is a serious or fatal injury, you must notify OSHA. The OSHA contact information needs to be present on the jobsite and at your main office, so make sure you have this posted in the job trailer and at your office. Failure to post this information is a violation, regardless of any injury.

Other Notices That Trigger OSHA Inspections

If OSHA receives a formal report or complaint of an unsafe condition at your jobsite, they will respond/investigate, usually within 3 days of getting the complaint.

If the complaint is by anonymous call or letter, OSHA can either investigate or it can send you a letter informing you of the complaint. Your attention to the issue should be immediate, and any corrective action that needs to be undertaken should be undertaken. In any event, your jobsite may be placed on a list for periodic inspection with particular attention paid to the issue in any complaint; anonymous or otherwise.

The OSHA Inspection

When OSHA shows up on the site, there are procedures that the inspectors must follow, and there are certain rights that you have and can assert.

One immediate right that you have, in the case of a periodic inspection, is the right to ask the inspector to obtain an administrative warrant. This will postpone the inspection, but it may also piss off the inspector. If there is no reason to put off the inspection, don't. Of course, if you let the inspection take place without a warrant, you have waived the lack of a warrant as a defense if the inspection results in citations and fines. In any event, be aware that you have due process rights even for this sort of administrative action, and you can assert those rights if you chose to.

When the OSHA inspector comes to the jobsite, if the managers of your company are at another location, it is a good idea to have the foreman/superintendent call them. If you feel comfortable with the jobsite team handling the inspection, this may not be necessary, but any citations and violations will be assessed against the company, not the jobsite team. It is probably a best practice to have the main office at least notified, and better yet, have someone with authority present for any OSHA inspection.

If you want someone from upper management present, nothing should happen on-site until they arrive. The inspectors should be made comfortable in the job trailer and not allowed to walk the site until you are ready. Again, cooperation is the order of the day and if your managers cannot get to the site in a timely manner, arranging for the inspection the next day, or at some other time, is not an inappropriate request.

The Inspection Protocol

There is a protocol that will be followed for any OSHA inspection.

First there will be an opening conference. The inspection team will present their credentials and explain the purpose of the visit, that is, the complaint that was submitted and/or the bases for the inspection. The inspection may, and likely will, include a walk around of the project site.

The scope of the inspection walk should be discussed and agreed on at the conference. You have a right to know what will happen and where the inspection will take place. OSHA has the right to inspect the entire site, provided it is safe and possible to do so.

The inspector will want to see your Injury and Illness Prevention Plan ("IIPP"), any MSDS documentation that you have assembled for that project site, and any HazMat information that may be available. This information should be present on the jobsite regardless, so make sure it is. Make sure that it is organized, and that the jobsite team understands what it means, where it is kept, and that it needs to be current and complete.

If the inspection team wants to talk to employees, the terms of any such discussions should be laid out at the opening conference. For example, does a Union rep need to be present, is the discussion going to be confidential and private, who can be present from the contractor's side, and will the inspector talk to individual workers during the walk or afterward? The parameters of the interview should be clear.

If the inspection will include a job site walk, the contractor representative can and should accompany the inspection team. OSHA inspectors can take pictures of the site, but they cannot take pictures of any proprietary or other privileged material. This should be discussed, and while it is unlikely that anything of a proprietary nature would be on a job site, there are those instances when the project means and methods might involve something unique that has some proprietary value. If so, OSHA does not get to take pictures, they can look, but they cannot photograph it.

If possible, the contractor should videotape the inspection. If the inspection team wants to break up into smaller groups and conduct simultaneous investigations, and assuming this is acceptable to the contractor, representatives for the contractor should accompany each OSHA team/inspector. Along with the video of the walk(s), or in lieu of it, you should take notes of what is done, said, looked at, touched etc. You need good notes, comprehensive notes, and readable notes; make sure that whoever is designated for this task can write clearly and can describe the inspection accurately. It's acceptable to make minor corrections for obvious problems at the time of the inspection. If you can, get them done and out of the way as you go.

OSHA can conduct tests and take samples during the inspection. Again, this should be discussed at the opening conference, because if OSHA intends to take samples or conduct tests, the contractor should have its own inspectors/experts there, as well. If nothing else, the contractor should take samples from the same locations as the inspector, and of course the location where any sample is collected should be identified and marked.

At the conclusion of the inspection, there will be an exit conference. The inspector(s) will review their findings with the contractor, including any comments or suggestions related to the IIPP, MSDS, HazMat or other matters. The inspector can issue subpoenas for witnesses at this time, and can issue citations for violations. They probably won't however.

Citations

OSHA has six months to issue citations after the inspection, so it is not likely that any citation will be provided at the exit conference, unless the conditions are so obviously dangerous that they pose an immediate threat to the workers health and safety. More likely, you will receive the citations later. The inspector has the right and the authority to require that certain evidence be preserved, especially if there has been an accident and an injury resulted. Make sure you comply. If you can't, or won't, make sure the inspector knows, and knows why.

Eventually, there will be a closing conference. This is where citations are typically issued. The types of citations that are issued have significant consequences, not just because the fines associated with the type of violation can be very substantial, but also because the OSHA violation figures into your Workers' Comp "mod rate" you are assigned by your Workers' Comp carrier, which in turn affects how much you pay for Workers' Comp insurance. The "mod rate" can impact your ability to pre-qualify for public works projects (and possibly private works projects too), so you should be very concerned about any OSHA citations that you receive.

Violations

There are five levels of OSHA violations: willful, repeat, serious, general, and regulatory. Willful is when the employer knows, or should know, the dangerous condition(s) exists but makes no effort to correct it. Repeat is a violation that was noted previously, was then corrected, but has within 3 years become an issue again. Serious is where there is a substantial probability of death or serious physical injury. General is one that affects the safety and health of the employee. Regulatory is just that, a technical violation, i.e. not posting permits or posters that may be required.

You do not want any OSHA violations on your record. Penalties for willful violations can be up to $70,000 for the first violation, but if the willful violation is a repeat violation, and if it causes a death, the penalty can be $100,000-$250,000 and from 1 to 4 years in jail. Serious violations can go as high as $25,000, and if it is a repeat violation, that penalty increases incrementally as well.

Even regulatory violations can be costly. The first violation can warrant a fine of $5,000, the 2d repeat violation of the same regulation will cost you $10,000 and the third will run you $20,000. There are better ways to spend this type of money, and, given that it is jobsite and worker safety that is at issue, there is no excuse for a violation of a regulation, let alone a repeat violation.

Multi Employer Jobsites

On multi-employer worksites, both construction and non-construction, citations may be issued only to the following categories of employers when the OSHA has evidence that an employee was exposed to a hazard in violation of any requirement enforceable by OSHA:

- The employer whose employees were exposed to the hazard (the exposing employer);
- The employer who actually created the hazard (the creating employer);

- The employer who was responsible, by contract or through actual practice, for safety and health conditions on the worksite; i.e., the employer who had the authority for ensuring that the hazardous condition is corrected (the controlling employer); or
- The employer who had the responsibility for actually correcting the hazard (the correcting employer).

Note: The employers listed may be cited regardless of whether their own employees were exposed to the hazard.

Right to Appeal the Citation

If you receive a citation, you have the right to appeal it. OSHA has six months to serve you with a citation after the close of the investigation. However, you, the contractor, have only 15 days from the date you receive the citation to notify the agency that you intend to appeal the citation(s). Your notice of appeal is made by filing a form or sending a letter to OSHA at the address provided on the citation or cover letter for the citation(s) that informs OSHA that you are going to appeal the citation(s).

There are four grounds for appeal:

- The existence of the violation and the how it is characterized; willful, serious, etc.
- The reasonableness of the time you are given to correct the alleged violation.
- The reasonableness of the actions you are required to take to make the corrections; and,
- The reasonableness of the penalty assessed.

There are also affirmative defenses that you can raise to the citation. Affirmative defenses are basically reasons that relieve you from any liability or responsibility for the violation. For OSHA citations, the affirmative defenses are limited but fairly comprehensive, and they can be quite effective.

The five affirmative defenses, that you must raise or else you waive them, are

- Independent employee action: you have an appropriate safety program in place and the employee knows it, you enforce it with specific penalties, but the employee acts contrary to it and is injured.
- Unforeseeable employee act: you have an appropriate safety program and the jobsite is properly supervised, the particular hazard is protected against if the rules are followed, but you did not and could not know of the hazard.
- Insufficient evidence that the employee was exposed to any hazardous condition: where OSHA cannot or has not shown that the injured employee was actually exposed to the condition claimed to exist.
- Vague, ambiguous, overbroad safety order: where it is not clear what the violation is, or what the statute requires to be done so that the contractor has to guess what it required; and
- Employer's lack of knowledge of the serious violation: where despite reasonable and proper efforts, the employer could not know there was a violation.

On multi-employer jobsites, there is a defense that the cited employer did not control the conditions or create the conditions that caused the injury, and/or that they took reasonable precautions to protect against just such a hazard. It may be that there was a violation, but another employer was responsible for the injury-causing circumstance, not you. In other words, it was not you or your operations that caused the injury.

Keep in mind that there is liability for failure to inform others of hazards that you know exist, even if they are not caused or created by you or your work. (24) If you or your employees know of a hazard, even if they did not create it, they must inform the project superintendent of the problem, or there could be a shared liability for the violation. If you, or your employees, know of a hazard and know that others may not be aware of the hazard, you and your people need to say something so that the hazard is abated. Safety is and should be a job-wide concern.

The Administrative Hearing

Assuming a citation is issued and an appeal is filed, the administrative hearing process is started. The cited contractor will have rights to discovery and to a hearing. The administrative hearing process is similar to but less formal than the judicial process. Rules of evidence are more relaxed in the administrative hearing than in trial, but are still present. The fact that it is a less formal process should not lull you into thinking it's something that does not need to be taken seriously. It is, and you must be prepared.

More often than not, the citation and penalty are negotiated and settled before any hearing. The contractor wants the violation "status" reduced, from willful to serious or, better yet, to general. The nature of the penalty is what is negotiated, and the various affirmative defenses are asserted to show that the cited contractor did not act in the manner that the citation contends. If the citation is reduced in degree, the penalties will likely also be reduced. The amount of the penalty is not what should drive the negotiation; it is the nature of the violation.

In this situation, where there is a citation that has significant consequences, it is advisable to retain a good attorney to defend you at the hearing, and to assist you in, if not actually prepare, the appeal, and represent you at the settlement negotiations. If the contractor has been really lax in its jobsite safety program, a good attorney is an absolute necessity, and if the contractor has been diligent in its safety program, a good attorney is likewise just as necessary.

Writ of Mandate and Judicial Recourse

If you go through the administrative appeals process and get an adverse decision, your next step is to seek a Writ of Mandate from the Superior Court that will direct the Appeals Board to change its decision in some way. If that is not issued, then the next step is to appeal to the Superior

Court and let the judge hear the matter again. If the Superior Court still gets it wrong, the matter goes to the Court of Appeals. It gets costly, but sometimes it is necessary.

A safe workplace avoids all of this, however. The job goes faster, the profit margin goes up, the general overhead costs come down, and it avoids the whole OSHA issue in the first place; safety first, as they say.

Conclusion

CONTRACTORS HAVE MADE A lot of money in the business, and they have done it with skill and integrity, for the most part. There are always exceptions to the rule, and sometimes those exceptions color the rule. The problem is that courts and juries often have the stereotypical expectation of the big galoot contractor whenever a contractor is involved in a dispute. The reputation precedes the reality, and so contractors often have an uphill battle making their case, no matter how compelling the facts are in their favor. As we said in the Introduction, it is often the case that the Red Queen has already decided the case before anyone says anything.

In many ways, this is the construction industry's own fault. As an industry, we have not always done a very good job of putting our best foot forward or, more important, educating the courts and the Legislature about what we do and why we do it. It is very easy for someone to pick apart any construction project and point out all the flaws in the finished product. Construction lends itself to that sort of hyper-critical review because any construction project is composed of any number of disparate and dissimilar parts that are assembled on-site in an uncontrolled environment and according to a plan that may, or may not, take into account even the most simple of facts; like the fact that the various pieces don't necessarily expand and contract at the same rate.

The rules are made by people who don't understand these limitations, and those rules are enforced by people who sometime know even less. The California Supreme Court has held, for example, that a window assembly (window, framing, flashing, glazing) are all part of a single unit, and that this unit is analogous to any other manufactured product. Consequently a window assembly is subject to the same strict liability analysis that would apply to determining whether, for example, a car is defective because some part goes wrong and causes an injury to someone or something.

If you start with that proposition, you are bound to find that contractors and construction generally is simply a matter of degrees of shoddiness. Cars and other manufactured goods are assembled in factories on assembly lines. There is a great deal of control that can be exercised over the manufacturing process in that environment; a level of control that is, for the most part, completely unavailable on a construction jobsite.

Factory workers don't get rained on, they don't run into unforeseen conditions like boulders in the way, and they don't need to deal with designs that are less than complete, and/or designers and owners whose first reactions to any question are defensive and querulous rather than cooperative and helpful. In short, construction is not manufacturing.

Yet that is exactly what the courts have come to understand and how they analogize the industry in the decisions that they hand down and impact the industry. How did we get to this state of affairs? Because the industry has done a really poor job of educating the courts and the public about what construction is and what contracting is all about. Not to mention that there are some really bad players in the industry, just like any other industry, and these bad players have caused the rest of the industry to be tarred with the same bad brush.

In order to point out that the bad actors are the exception and not the rule, as far as what contractors do, the industry needs to both police itself and also do better at educating the judges and legislatures that make the rules. We spoke about SB 800, the Right to Repair Act, which was passed and (supposedly) corrects some of the injustice that ran rampant in the home-building industry for so long. Whether you agree that this is a good piece of legislation or not, the fact is that it is legislation, it was fostered by the construction industry, and it did address some of the problems that exist in home building.

As an industry, we need more than this, and the way to get more, the only way, is for contractors to get together and push for changes. Pushing for change doesn't necessarily always mean going to Sacramento and glad-handing a bunch of good old boys and girls so that they will listen to your spiel about what change you want put on the books.

One way to implement change is by getting together with like-minded contractors and industry professionals, and discussing the issues that are confronting your business on a daily basis. Identifying the problem is often just as difficult, but far more important, than identifying a solution. There are any number of builders' associations and industry groups, and these are the basis for any fundamental change in the impression of the industry and in the law that governs the industry. Put a bunch of builders associations together, and you have an even bigger group that has even more clout and influence.

Inviting judges or legislators to conferences to present talks, or to be part of panel discussions that address what the problems are that they see, from their perspective, would also provide an opportunity to both understand and explain why those perspectives are what they are and/or why they may not be wholly correct. We have two ears and one mouth, so we should listen twice as much as we talk, or so the old saying goes.

The point is to educate the judges and the lawmakers about the problems in the business and why, when they make decisions or laws, those decisions and laws have the consequences they do. Getting the Supreme Court to understand that tract home construction is not analogous to car manufacturing would be a great step forward.

Certainly there is no excuse for poor quality in home building, especially given the money spent constructing those homes. But the poor construction does not go away by pretending that the window leak is somehow the same as a faulty bumper or airbag. The rules that come from this sort of logic increase the cost of construction as the attorneys step in and try and exploit the new case law, and thereby create even more new litigation. With every new case involving construction, the non-construction public is reinforced in their preconceived notions that contractors are just a bunch of money-hungry scallywags who don't care about anyone but themselves.

This is not to say that the law is not changing. It is. The courts are reining in public agencies, for example, and not allowing them to circumvent the rules and avoid the law because they have the money to carry on a fight and/or can bankrupt any but the most well-heeled contractors if that contractor has the audacity to dispute the agency's tactics. For example, recent decisions have required public agencies to disclose their criteria for determining how they will decide best value in a bid process before the contractors actually submit their bid and, as another example, to make post-bid information available to all bidders, not just the favorite bidder.

Public agencies have also been called to task for spending tens of millions of dollars pursuing claims against contractors, when those claims could have been settled for a fraction of that amount. Especially in this economic environment, it makes no sense to the taxpayer (or anyone, for that matter) that personal vendettas are driving lawsuits against contractors and that the funding for that lawsuit is coming at the expense of schools and closed life-safety facilities. There is a limit to the stupidity of public officials that the public will tolerate, but whether they tolerate it or not, the construction industry's reputation is also tarnished.

We can and should do better. The courts seem to be catching on to certain abuses that are prevalent in the industry, but they don't yet understand all they should about the construction industry and how it functions. Shame on us; we should do better. Get together with your local Builders Association and have them invite a judge from the local court to address the Association on issues that the court is seeing as part of its local docket. Round-table the issues, ask good questions, and educate the judge in the course of doing it.

OK, so enough. We are out of the rabbit hole and back in the real world...

You, as the contractor, will be building your project, and some of the concerns and issues raised in this book will come up; but hopefully only a few. If they do, this book hopefully provides direction about the best way to proceed and what to do in the situation. If the circumstance you are confronted with is not addressed, or you are not sure if it is addressed, then the book should provide you with some basic legalese sufficient to let you have at least a reasonable discussion with an attorney, because that should be your next step.

A couple of things to consider if you do need to call an attorney.

First, make sure they practice construction law. You don't want someone that does family law to handle your construction dispute. Similarly, you probably want an attorney who has a practice that includes construction law, but also some of the related fields, like real property, corporations, business law, employment law; those sorts of things. An attorney with a practice in areas of law that would include issues related to the construction industry is better, all things being equal, than someone who has an esoteric practice that includes a lot of disparate areas that may not overlap.

You should also ask around for attorney recommendations. If nothing else, make sure the attorney is someone that you can work with. Go with your gut; if you don't like the guy when you meet him, you probably won't like him when you work with him. Find out his hourly rate, too. If your issues are simple and fairly common, you likely don't need the super-high-powered attorney from the big city law firm representing you.

Last but not least, understand that in California it is mandatory that you and your attorney have a written agreement for services. It is a contract, and, just like all the other contracts we have discussed here, you need to read it. It will explain the hourly rates, the retainer amount, the billing cycle, the dispute resolution procedure, and the scope of the representation. This last part is important. Make sure that you understand what the attorney understands is his/her assignment. If it is for one issue only, then that is all the attorney will be doing. If you need more, you may need to amend or enter a new agreement. Make sure that this part is very clear so there is no misunderstanding.

Now that you know so much, go build something to get rid of the headache.

Good luck and have fun!

Addenda: July 2012 Changes

There are a number of important changes that will occur and that have occurred but that have not been fully implemented as this book is written. The following is an outline of those new requirements and you should take note of them, especially the changes to the Conditional and Unconditional lien releases that take place after July 2012. Lien Releases are critically important and the language contained in the lien release itself is statutory, so be aware of the change.

Preliminary Notice On private works the General must give notice via the 20-Day Preliminary Notice to the construction lender. The name has also been changed. It is no longer a 20-Day Preliminary Notice; it is just a Preliminary Notice.

Mechanic Liens The Legislature undertook to simplify, or at least organize in a better way, the requirements for Mechanic Liens. These requirements were found in Civil Code section 3082-3267 but will now be located in various statutes in the 8000 section. Whether this makes it easier or more confusing remains to be seen but the changes are important.

Lien Release Language The lien release language has been completely redone and must be used in substantially the same form as it is provided in Civil Code section 8132. Like its predecessor statute- Civil Code section 3262- if the lien release form you use does not have the same or substantially the same language it is ineffective to release the lien right even if you pay the amount and the funds are deposited. The new language is included in the Appendix at the end of the book. This change only takes place as of July 1, 2012; until then the language in section 3262 is the correct release language to use.

Notice of Mechanic Lien Recall that before you can record a Mechanic Lien you need to provide notice to the Owner that this is in the offing if you are not paid. Now, you will need to file a Proof of Service for the Notice of Mechanic Lien when you file and serve your lawsuit otherwise the Mechanic Lien is invalid. Remember to provide the Notice and provide a Proof of Service too so that you can attach it to your pleading.

Expunging Mechanic Liens- If you file a Mechanic Lien that is improper and the Owner goes to court to have it expunged (removed) from title, the attorney fees that can be awarded are no longer limited to $2000. Instead, the Owner can obtain all "reasonable' attorney fees. Bottom line, watch your liens and make sure they are valid.

Notice of Completion A couple of new things here. First, the time for the owner to record the notice of completion on private works is extended from 10 days to 15 days. Second, the 'acceptance by the owner' is not a valid form of "completion" any more. There is still the beneficial occupancy and cessation of labor as the equivalent of completion. Finally, on public works, the cessation of work that equates to completion is 60 days, the same as private works.

Construction Loans The lender must now be identified on any contract between owner and general (direct) contractor and this information has to be provided to any contractor/supplier that provides a Preliminary Notice.

Licensing Law The Legislature has approved LLC's as valid entity forms to obtain contractor's licenses. The change was authorized in 2011 but the finer parts of the requirements were being worked on. If you want to have your LLC become a licensed contractor, then you will need to post and maintain a surety bond in the amount of $100,000 for employees that may not be paid and you will need to obtain and provide proof of errors and omissions insurance for the LLC members/employees in the amount of $1M or $5M depending on the number of LLC employees/members.

Workers' Compensation These changes are already in effect; you need to have Workers' Comp if you have employees. If you do not and you are working and have employees, and you are caught, the job will get shut down, you may be fined, and you could be jailed for up to 60 days on the misdemeanor charge that now attaches. Don't have employees if you don't have the Workers' Comp to cover them.

Other Changes The Legislature changed requirements for design professional liens and for landscape architects. If you do this kind of work check out the changes in the section on Mechanic Liens generally for design professionals and for landscape architects, who are allowed Mechanic Liens under new section 8014.

Endnotes

IF YOU DECIDE TO read the actual cases that are referenced, or any cases that are cited within a case, there are a couple of things to understand about the references. First, Cal. App. refers to California Appellate decisions. These are one of the reporters where appellate decisions are published. The other reference is Cal. Rptr. (California Reporter) or Pac. Rptr. (Pacific Reporter). The most common reference is the Cal. App. series.

When you refer to any appellate reporter, you will see a volume number before the name of the reporter and then the edition number; i.e.- 24 Cal App 3d. Make sure that you are looking at the correct edition. We are now at the 4th edition for the Cal. App series. The California Reporter is at the 3d edition and Pacific Reporter is also at the 3d. [Pacific Reporter is a regional reporter, not just California.] As you can see, they are all different in their respective volume numbers. The edition number will be followed by the page number in the volume where the particular case will be found.

When you are looking for cases, be aware that there are Cal. editions too. These volumes are cited as i.e.- 24 Cal 4th with the page number following. These volumes are California Supreme Court cases. Be sure you are not looking for an appellate case (Cal. App.) in the Supreme Court reporters (Cal.). It is easy to be confused.

Finally, one last caution, the pagination in the Cal. App. series is not the same as the Cal. Rptr. or Pacific Rptr. series. If you are reading the decisions you may find a bunch of bracketed numbers that indicate the page number for the different series of reporters. You will need to check the beginning page number for the case so that you can coordinate the page with the reporter as you read it and especially if you want to cite the case to the court for some reason. Make sure you give the correct page number and the correct reporter so the judge can find the point that you want to make.

Enjoy!

1. *WSS Industrial Construction, Inc. v. Great West Contractors, Inc.* (2008) 162 Cal. App. 4th 581: WSS contends it substantially complied with the licensing requirements because, as the corporation's RMO, Ramirez previously qualified a WSS partnership for a contractor's license and held various individual contractor licenses of his own at all times before and after the corporation obtained its license. WSS has not made a sufficient showing of substantial compliance. The contract at issue was between Great West and WSS, the corporation, the entity which bid the project and the only entity to have performed work for which recovery was sought in this action, not WSS, the partnership or Ramirez as an individual. WSS was incorporated for almost two years before it applied for a contractor's license in August 2001. Unequivocal evidence at trial showed the sole license held by the corporation "was issued 12/21/2001 and did not exist before that date." The WSS partnership license was not — and could not have been — transferred or reissued to WSS, the corporation. (See § 7075.1, subd. (a) ["No license ... shall be transferable to any other person or entity under any circumstances."].) The licensing history of the partnership and Ramirez are not relevant to the WSS corporate license history.

   Opp v St. Paul Fire & Marine Ins. Co (2007) 154 Cal. App. 4th 71: The trial court ruled that a California licensed building contractor could not recover against a surety in an action on a payment bond. In his capacity as president of a corporation that was not licensed in California, the contractor entered into a subcontract for work in California, using his individual contractor's license number. The trial court ruled that he was not a party to the contract and that the corporation could not recover on the payment bond because it was unlicensed.

2. *White v Cridlebaugh* (2009) 178 Cal. App. 4th 506: The court held that the contractor violated California's licensing requirements and because of that violation, the homeowners were entitled to recover all compensation paid to the contractor for the unlicensed work. The corporation was not qualified for a contractor's license because the individual who was identified as the qualifier was not actively engaged in its construction business and no replacement was ever qualified. Therefore, the corporation's license was suspended by operation of law.

3. *Vallejo Development Co. v Beck Development Co.* (1994) 24 Cal. App 4th 929: By entering into the agreements to "improve the Property" and to be "solely responsible for completion of" infrastructure improvements--including graded building pads, storm drains, sanitary systems, streets, sidewalks, curbs, gutters, utilities, street lighting, and traffic signals--VDC was clearly contracting to provide construction services in exchange for cash payments by respondents. The mere execution of such a contract is an act "in the capacity of a contractor," and an unlicensed person is barred by section 7031, subdivision (a), from bringing claims based on the contract. Even if VDC performed only administrative and oversight functions with respect to the actual installation of infrastructure improvements, it nevertheless acted "in the capacity of" a general engineering contractor by performing those functions in fulfillment of contractual obligations owed to the owners of the property on which the improvements were installed. The fact that VDC subcontracted with licensed contractors to provide the actual labor, equipment and materials to construct the infrastructure improvements is irrelevant. Section 7026 plainly states that both the person who provides construction services himself and one who does so "through others" qualifies as a "contractor." The California courts have also long held that those who enter into construction contracts must be licensed, even when they themselves do not do the actual work under the contract.

4. *Wright v Issak* (2007) 149 Cal. App. 4th 1116: In relevant summary, section 7125.2 states that a contractor's license is automatically suspended as of the date the contractor was required to obtain workers' compensation insurance but did not. This language cannot be clearer. Its effect is that, because plaintiff underreported his payroll and, thus, did not obtain workers' compensation insurance in 2004, plaintiff's license was suspended before, during, and after he performed work on defendants' home.

5. Substantial Compliance Cal Buss & Prof Code § 7031:

   (b) Except as provided in subdivision (e), a person who utilizes the services of an unlicensed contractor may bring an action in any court of competent jurisdiction in this state to recover all compensation paid to the unlicensed contractor for performance of any act or contract. (e) The judicial doctrine of substantial compliance shall not apply under this section where the person who engaged in the business or acted in the capacity of a contractor has never been a duly licensed contractor in this state. However, notwithstanding subdivision (b) of Section 143, the court may determine that there has been substantial compliance with licensure requirements under this section if it is shown at an evidentiary hearing that the person who engaged in the business or acted in the capacity of a contractor (1) had been duly licensed as a contractor in this state prior to the performance of the act or contract, (2) acted reasonably and in good faith to maintain proper licensure, (3) did not know or reasonably should not have known that he or she was not duly licensed when performance of the act or contract commenced, and (4) acted promptly and in good faith to reinstate his or her license upon learning it was invalid.

6. *MW Erectors Inc. v Niederhauser Ornamental and Metal Works Co Inc* (2005) 36 Cal. 4th 412: The provisions make clear the general rule denying recovery of all compensation for work requiring a con-

tractor's license if a valid license was not in place when performance began, or if licensure lapsed at any time during the work. That statute prohibits a contractor from suing for the collection of compensation for the performance of any act or contract where a license is required ... without alleging that he or she was a duly licensed contractor at all times during the performance of that act or contract. The "act" of executing an agreement is not one for which a contractor seeks compensation; rather, he or she pursues payment for carrying out the contract in a satisfactory manner. Hence, we conclude, the due licensure of which section 7031(a) speaks is due licensure while the contract itself is being performed.

7. *Pacific Caisson & Shoring v. Bernards Bros.,*(2011) 2011 Cal. App. LEXIS 1090: Section 7031, subdivision (a) reads in relevant part: "Except as provided in subdivision (e), no person engaged in the business or acting in the capacity of a contractor, may bring or maintain any action, or recover in law or equity in any action, in any court of this state for the collection of compensation for the performance of any act or contract where a license is required by this chapter without alleging that he or she was a duly licensed contractor at all times during the performance of that act or contract, regardless of the merits of the cause of action brought by the person ... ."

8. *Great Western Contractors v WSS Construction* (2008) 162 Cal. App. 4th 581: WSS submitted a second invoice for $ 11,000 on December 19, 2001, "for work completed to date," which included $ 9,000 for shop drawings and $ 2,000 for specialized anchor bolts WSS ordered and had delivered to the project. Ramirez testified that ordering materials — such as anchor bolts — was included in WSS's subcontract. He also testified he knew licensure was required for preparation of shop drawings, and that he knew WSS did not have its contractor's license when it submitted its first two invoices. In order to bill for work provided under contract a contractor's license is required at all times.

9. *Holland v Morse Diesel Internat., Inc.* (2001) 86 Cal. App. 4th 1443: Causes of action that do not seek "the collection of compensation for the performance of any act or contract for which a license is required" are beyond the scope of Business and Professions Code section 7031. Holland's claims of racial discrimination in the form of derogatory comments on the jobsite and retaliation for complaining about those comments are "external to the arrangement for construction work as such, and ... thus unrelated to any protective concern of the licensing law." (*Hydrotech*, supra, 52 Cal. 3d at p. 1001.) The same cannot be said, however, for Holland's allegations that he was not properly compensated, as Caucasian contractors were. These claims directly implicate the statutory proscription against recovery of compensation by unlicensed contractors. Nevertheless, they raise a question of appropriate remedy, not a bar to maintenance of a civil rights cause of action.

10. *Owner Builder* exception: Section 7044 of the CSLL provides that the prohibition against an unlicensed contractor from bringing or maintaining an action to recover compensation do not apply to (a) an owner of property, building or improving structures thereon, who does the work himself or herself or through his or her own employees with wages as their sole compensation, provided none of the structures are intended or offered for sale; or (b) an owner of property, building or improving structures thereof, who contracts for such a project with a subcontractor or subcontractors thereon; or (c) a homeowner improving his or her principal place of residence, providing that (1) the work is performed prior to sale; (2) the homeowner has actually resided in the residence for the 12 months prior to completion of the work; and (3) the homeowner has not availed himself or herself of the exemption on more than two structures more than once during any three year period.

11. Buss & Prof Code § 7114. Aiding, abetting, or conspiring with unlicensed person to evade law:

(a) Aiding or abetting an unlicensed person to evade the provisions of this chapter or combining or conspiring with an unlicensed person, or allowing one's license to be used by an unlicensed person, or acting as agent or partner or associate, or otherwise, of an unlicensed person with the intent to evade the provisions of this chapter constitutes a cause for disciplinary action.

(b) A licensee who is found by the registrar to have violated subdivision (a) shall, in accordance with the provisions of this article, be subject to the registrar's authority pursuant to Section 7099 to order payment of a specified sum to an injured party, including, but not limited to, payment for any injury resulting from the acts of the unlicensed person.

12. The Business & Professions Code needs to be read in multiple parts for issues related to unpaid judgments and license suspension.

Section 7090.1. Automatic suspension of license for failure to pay civil penalty or comply with order of correction; Contest of determination; reinstatement; Delay in revocation for good cause; Dissociation from other licensee

(a)(1) Notwithstanding any other provisions of law, the failure to pay a civil penalty, or to comply with an order of correction or an order to pay a specified sum to an injured party in lieu of correction once the order has become final, shall result in the automatic suspension of a license by operation of law 30 days after noncompliance with the terms of the order.

(b) The cited licensee shall also be automatically prohibited from serving as an officer, director, associate, partner, manager, or qualifying individual of another licensee, for the period determined by the registrar, and the employment, election, or association of that person by a licensee shall constitute grounds for disciplinary action.

Section 7071.17. Bond required where applicant or licensee has unsatisfied final judgment for failure to pay contractor, subcontractor, consumer, materials supplier, or employee; Notice of unsatisfied final judgments; Suspension for noncompliance; Reinstatement; Disqualification from serving as personnel of record for other licensee; Disciplinary action

(a) Notwithstanding any other provision of law, the board shall require, as a condition precedent to accepting an application for licensure, renewal, reinstatement, or to change officers or other personnel of record, that an applicant, previously found to have failed or refused to pay a contractor, subcontractor, consumer, materials supplier, or employee based on an unsatisfied final judgment, file or have on file with the board a bond sufficient to guarantee payment of an amount equal to the unsatisfied final judgment or judgments. The applicant shall have 90 days from the date of notification by the board to file the bond or the application shall become void and the applicant shall reapply for issuance, reinstatement, or reactivation of a license. The board may not issue, reinstate, or reactivate a license until the bond is filed with the board. The bond required by this section is in addition to the contractor's bond.

[...]

(g) Except as otherwise provided, the bond shall remain in full force in the amount posted until the entire debt is satisfied. If, at the time of renewal, the licensee submits proof of partial satisfaction of the financial obligations covered by this section, the board may authorize the bond to be reduced to the amount of the unsatisfied portion of the outstanding judgment. When the licensee submits proof of satisfaction of all debts, the bond requirement may be removed.

13. *Crawford v Weather Shield Mfg. Inc* (2008) 44 Cal. 4th 541: The subcontract at issue included a further express indication that the express duty "to defend" actions against JMP was not strictly limited to those claims on which, in the end, Weather Shield actually owed indemnity. The indemnity and defense clauses of the subcontract

contained linguistic differences that conform to the logical distinctions between the two duties.

On the one hand, the subcontract obligated Weather Shield to "indemnify ... [JMP] ... against" all claims for injury, damage, loss, or theft arising from performance of the subcontract, while, on the other, it required Weather Shield "to defend any suit or action ... against [JMP] founded upon the claim" of such injury, damage, loss, or theft. (Italics added.) One can only indemnify against "claims for damages" that have been resolved against the indemnitee, i.e., those as to which the indemnitee has actually sustained liability or paid damages. Indemnification, af.ter all, is the act of saving another from the legal consequence of an act. (§ 2772.) Hence, a clause requiring Weather Shield to indemnify JMP "against" defined claims clearly indicated that the indemnity obligation would apply only if JMP ultimately incurred such a legal consequence as a result of covered claims.

By contrast, as noted above, the subcontract required Weather Shield "to defend" JMP against "any suit or action ... founded upon the claim of such damage ... ." (Italics added.) Under this language, the duty to defend arose, as it logically must, as soon as a "suit or action" was brought against JMP that was "founded upon" a covered claim, i.e., that asserted a claim within the coverage of both clauses. Necessarily, a duty expressed in this manner did not require a final determination of the issues, including the issue of Weather Shield's negligence, before Weather Shield was required to mount and finance a defense on JMP's behalf.

14. *Michealis v Benavides* (1998) 61 Cal App. 4th 681: Directors and officers are not personally liable on contracts signed by them for and on behalf of the corporation unless they purport to bind themselves individually. ... [P] Directors or officers of a corporation do not incur personal liability for torts of the corporation merely by reason of their official position, unless they participate in the wrong or authorize or direct that it be done. They may be liable, under the rules of tort and agency, for tortious acts committed on behalf of

the corporation. They are not responsible to third persons for negligence amounting merely to nonfeasance, to a breach of duty owing to the corporation alone; the act must also constitute a breach of duty owed to the third person. More must be shown than breach of the officer's duty to his corporation to impose personal liability to a third person upon him."

15. *Schram Construction, Inc. v. Regents of University of California* (2010) 187 Cal. App. 4th 1040: On its face, the statute mandates publication of the procedures and criteria necessary to "ensure that all selections are conducted in a fair and impartial manner." (§ 10506.4, subd. (c).) It follows that information regarding the procedure and criteria for selecting bid packages (bid package selection information) must be disclosed to the extent the bid selection would otherwise be unfair or its integrity would be compromised by nondisclosure. We conclude publication of the bid package selection information was necessary to a fair and impartial bid selection in this case.... A contractor is placed at a competitive disadvantage, not only when undisclosed information undermines its ability to bid on a common basis, but also when the nondisclosure is material in deciding whether to submit a bid in the first instance.

16. *Burchett v. City of Newport Beach* (1995) 33 Cal.App.4th 1472 (also *Katsura v. City of San Buenaventura*, 2007 Cal. App. Unpub. LEXIS 6583): Persons dealing with a public agency are presumed to know the law with respect to any agency's authority to contract. "'One who deals with the public officer stands presumptively charged with a full knowledge of that officer's powers, and is bound at his ... peril to ascertain the extent of his ... powers to bind the government for which he ... is an officer, and any act of an officer to be valid must find express authority in the law or be necessarily incidental to a power expressly granted.'"

17. *Arntz Builders v Sup Ct* (2004) 122 Cal App 4th 1195: The public importance of ensuring the complete impartiality, and public con-

fidence in the impartiality, of the courts is hardly less compelling than is true for the system of arbitration. As we have long held, the purpose underlying the mandatory change of venue provision in section 394 "is to guard against local prejudices which sometimes exist in favor of litigants within a county as against those from without and to secure to both parties to a suit a trial upon neutral grounds." Regardless of whether the danger of actual prejudice is sufficient to warrant transfer under section 397, section 394 "is designed to obviate the appearance of prejudice as well as actual prejudice or bias." Thus, the public purpose that underlies section 394 provides an additional and particularly compelling reason why the rights it confers cannot be waived by advance agreement of the parties. This consideration is over and above the more general legislative concern that parties not "disrupt the statutory [venue] scheme and bring the administration of justice into disrepute in order to have their cause heard where they believe it will be received most sympathetically." Because we conclude that the provision in the construction contract between Arntz and County purportedly waiving the benefits of section 394 is void, and because section 394 affords Arntz the right to transfer the action to a neutral county, we need not address whether transfer was required by section 397.

18. *Sehulster Tunnels/PreCon v Traylor Bros Inc/Obayashi Corp* (2003) 111 CalApp 4th 1328: California law protects the interests of the subcontractor by providing that a subcontractor's claim "passes through" the General Contractor to the owner. Consequently, by contract or settlement agreement, a General Contractor and a subcontractor can agree that the contractor will pass through the subcontractor's claims against the contractor to the owner. When a public agency breaches a construction contract with a contractor, damage often ensues to a subcontractor. In such a situation, the subcontractor may not have legal standing to assert a claim directly against the public agency due to a lack of privity of contract, but may assert a claim against the General Contractor. In such a case,

a General Contractor is permitted to present a pass-through claim on behalf of the subcontractor against the public agency. The pass-through process and such agreements are designed to shorten the legal process by not requiring the subcontractor to first sue the General Contractor and the latter to actually pay damages to the former before suing the owner.

The record clearly shows that all parties were fully aware that it was Sehulster's claim TBO was passing through. TBO, like City, believed Sehulster's claim to be meritless. Consequently, TBO neither challenged City's rejection of its pass-through request nor sued City for breach of contract.

City characterizes Sehulster's pass-through claim in effect as TBO's claim, and contends that TBO had the option of contesting City's decision denying the claim and asserting it was entitled to receive the additional $ 2,545,000 for breach of contract. City points out that had TBO done so, it would have been required under section 16.4.2.1 of the Prime Contract to submit "its" claim to the DRB. Instead, City contends that TBO agreed that the claim lacked merit and thus admitted it did not have a viable claim against City for breach of contract. TBO did not present "its own claim" against City, and City is correct that TBO agreed the design changes would not affect the negotiated price of the Prime Contract and that Sehulster's claim was unfounded. [TBO therefore had the obligation to pay the judgment that was awarded Sehulster Tunnel following trial.]

19. *Geo. Hyman Constr. Co. v United States* 30 Fed Cl. 170 (Fed Cl 1993): When the subcontract contains a clause completely exonerating a prime contractor from liability to its subcontractor for the damage complained of, suit cannot be maintained by the prime contractor against the Government," This statement cannot be accommodated to a conclusion that such a suit may be maintained when the subcontractor completely exonerates the prime contractor by modification of the subcontract, i.e., by a release. Permitting the parties subsequently to "revive" (or to create a new) liability by

mutual agreement would essentially nullify the Severin doctrine in every action before this court, since the very fact that the prime contractor is pursuing the subcontractor's claim evidences that just such an agreement exists. Permitting the contractor and subcontractor to create an obligation solely for purposes of suit against the government by an agreement entered into subsequent to the government contract and subcontract would vitiate every purpose of the *Severin* doctrine to prevent assignment of claims against the government, to prohibit imposition of liability against the government without its consent, and to prevent actions against the government by those lacking privity of contract with the government.

20. *Westinghouse Electric Corp. v County of Los Angeles* (1982) 129 Cal. App. 3d 771: Whether a stop notice claimant, who is not a party to the prime contract, has standing to challenge a liquidated damages assessment which has reduced the contract retention fund available to satisfy the stop notice, appears to be a question of first impression. However, the short answer, "yes, the claimant has standing," flows from Civil Code section 3186. That section requires the public entity to withhold from the original contractor money due or to become due. The stop notice affects what has variously been called an equitable garnishment or an assignment of that money. Funds withheld from the contractor as liquidated damages are not "earned" by the contractor. Hence, they are not "due or to become due." Since the stop notice claimant has an interest, not merely in whatever fund the public entity and the prime contractor agree has been properly retained, but in all money due or to become due (up to the amount of the claim), the stop notice claimant has a right to assert that more is due than the contracting parties were willing to leave in the retention fund. "Once the stop notices were filed properly ... and there was at the time of filing or thereafter money earned by the contractor and unpaid, an equitable assignment or garnishment pro tanto in favor of the stop claimants occurred and a cause of action thereon arose in their favor." Furthermore, the stop notice

claimant, in a sense, stands in the shoes of the contractor with reference to the withheld funds. The contractor could have denied the validity of the liquidated damages provision and maintained the fund intact, and so may Wesco (and its assignee, Sherwin).

21. *Liton Gen. Engineering Contractor, Inc. v. United Pacific Insurance* (1993) 16 Cal. App. 4th 577: Liability for attorney fees was statutorily based, and did not depend upon any contractual liability for attorney fees by the principal. To interpret the relevant statutes otherwise would undermine the mandatory fee shifting provisions of Civil Code sections 3248 and 3250. Similarly, to conclude that a surety may escape the statutory duty to pay fees in an action on a payment bond when the principal finally pays the amount due to the subcontractor, following protracted and expensive litigation-related proceedings, would also subvert the fee shifting statutes. Such result would make the statutory mandate dependent upon (1) whether the principal was contractually liable to the subcontractor for fees; and (2) whether the principal paid the underlying debt promptly following judgment, regardless of the time, energy and money expended by the subcontractor in obtaining judgment in its action on the bond. To condition application of section 3250 upon either occurrence as a practical matter would vitiate the strong public policy that informs the public works fee shifting statutes. The fee obligation under Civil Code section 3250 is a separate and primary obligation of the party (surety or claimant) who does not prevail in the action on the bond. Payment of the initial obligation does not exonerate the surety.

22. *North Bay Construction Inc v City of Petaluma* (2006) 143 Cal. App. 4th 552:: There is no right to impose a lien on property owned by a public entity unless such a right has been expressly conferred by statute. "One cannot sue the state, unless expressly authorized by statute, and this principle is embodied in our constitution. While sovereign immunity against being sued has been waived by Government Code section 945 (a "public entity may sue and be sued"), Government Code section 944 provides that "[n]othing in this part imposes li-

ability upon a public entity unless such liability otherwise exists." No statute imposes liability on a public entity for debts incurred by a lessee for improving property owned by and leased from the public entity. Nor can section 3109 be construed to implicitly authorize a mechanic's lien against public property for performing work that is not a "public work." To the contrary, the legislation adopting section 3109 provided expressly that it was the intent of the Legislature that "[t]his act shall not be construed to constitute a change in, but shall be construed as declaratory of, the preexisting law."

23. *Hill v. City of Clovis* (1998) 63 Cal. App. 4th 434: A party's voluntary dismissal without prejudice does not come equipped by law with an automatic tolling or waiver of all relevant limitations periods; instead, such a dismissal includes the very real risk that an applicable statute of limitations will run before the party is in a position to renew the dismissed cause of action. Also, a voluntary dismissal does not protect a cross-complainant from a later contention that a dismissed cause of action in a cross-complaint was compulsory and therefore required to be brought and adjudicated in the action initiated by the plaintiff. Instead, such a dismissal raises the possibility that the judgment in the action, if affirmed on appeal, will activate the statutory bar against any subsequent attempt by the cross-complainant to reinitiate prosecution of the dismissed cause of action.

[see also *Wells v. Marina City Properties, Inc.* (1981), 29 Cal. 3d 781: The implication of a subdivision 1 dismissal is that it is generally "without prejudice" to a plaintiff's filing of a new action on the same allegations if it is done within the period of the appropriate statute of limitations.]

NOTE: For Public Works: Be careful here because the time to file a lawsuit against a public agency is strictly controlled by statute and if you dismiss and refile, the statute of limitations may run as of the date of the dismissal See: *Martell v Antelope Valley Hosp. Medical Center* (1998) 67 Cal. App. 4th 978:: Here, appellants dismissed their first complaint despite section 945.6's language that "any suit" must be filed within six months of the public entity's rejection of a claim. Now, they seek relief from the consequences of their decision, but in doing so ignore that simply because "equity abhors a forfeiture[, it] does not excuse [them] from complying with the law."... Equitable tolling did not apply. But, even assuming it did, it does not help appellants because the Act's six-month statute of limitations would have been tolled only for the twenty-seven months appellants' first complaint was pending before its voluntary dismissal in May 1992. After May 1992, the six-month period would have started to run again and would have long since expired by the time appellants filed their second complaint almost five years later.

24. *Suarez v Pacific Northstar Mechanical Inc.* (2009) 180 Cal. App. 4th 430:: We agree with appellants' interpretation of section 6400 and regulation 336.11. Read together, and in light of the legislative history of the amendments to [Labor Code] sections 6304.5 and 6400, these two Cal-OSHA provisions impose a duty on each employer, at a multiemployer worksite, to report all nonobvious hazards about which the employer learns because its employees were exposed to them during the course of their work, even if the employer in question did not create the hazard. Moreover, a breach of that duty is actionable in tort by any worker at the site who is subsequently injured by the hazard that was not reported.

25. *Martinez v Brownco Const Co.* (2012) 203 Cal. App. 4th 507 : Where a party makes two section 998 offers to compromise more than 30 days apart, the purpose of section 998 is adequately served by the statute's existing language, which entitles an offeror to cost shifting from the date of the earliest reasonable offer. This rule encourages early settlement, respects the Legislature's preference for early discovery and evaluation of the merits of lawsuits, and presents a certain, bright-line rule by which settlement negotiations may be guided. If any mischief or confusion results from later offers, or any gamesmanship arises, the court can address such concerns when it awards costs.

Appendices

Appendix A

Use After July 2012

CIVIL CODE SECTION 8202

(a) The preliminary notice shall comply with the requirements of Section 8102, and shall also include:

(1) A general description of the work to be provided.

(2) An estimate of the total price of the work provided and to be provided.

(3) The following statement in boldface type:

NOTICE TO PROPERTY OWNER

EVEN THOUGH YOU HAVE PAID YOUR CONTRACTOR IN FULL, if the person or firm that has given you this notice is not paid in full for labor, service, equipment, or material provided or to be provided to your construction project, a lien may be placed on your property. Foreclosure of the lien may lead to loss of all or part of your property. You may wish to protect yourself against this by (1) requiring your contractor to provide a signed release by the person or firm that has given you this notice before making payment to your contractor, or (2) any other method that is appropriate under the circumstances.

This notice is required by law to be served by the undersigned as a statement of your legal rights. This notice is not intended to reflect upon the financial condition of the contractor or the person employed by you on the construction project.

If you record a notice of cessation or completion of your construction project, you must within 10 days after recording, send a copy of the notice of completion to your contractor and the person or firm that has given you this notice. The notice must be sent by registered or certified mail. Failure to send the notice will extend the deadline to record a claim of lien. You are not required to send the notice if you are a residential homeowner of a dwelling containing four or fewer units.

(b) If preliminary notice is given by a subcontractor that has not paid all compensation due to a laborer, the notice shall include the name and address of the laborer and any person or entity described in subdivision (b) of Section 8024 to which payments are due.

(c) If an invoice for material or certified payroll contains the information required by this section and Section 8102, a copy of the invoice or payroll, given in compliance with the requirements of Chapter 2 (commencing with Section 8100) of Title 1, is sufficient.

Appendix B

Use After July 2012

Civil Code Section 8132

If a claimant is required to execute a waiver and release in exchange for, or in order to induce payment of, a progress payment and the claimant is not, in fact, paid in exchange for the waiver and release or a single payee check or joint payee check is given in exchange for the waiver and release, the waiver and release shall be null, void, and unenforceable unless it is in substantially the following form:

CONDITIONAL WAIVER AND RELEASE ON PROGRESS PAYMENT

NOTICE: THIS DOCUMENT WAIVES THE CLAIMANT'S LIEN, STOP PAYMENT NOTICE, AND PAYMENT BOND RIGHTS EFFECTIVE ON RECEIPT OF PAYMENT. A PERSON SHOULD NOT RELY ON THIS DOCUMENT UNLESS SATISFIED THAT THE CLAIMANT HAS RECEIVED PAYMENT.

Identifying Information

Name of Claimant:

Name of Customer:

Job Location:

Owner:

Through Date:

CONDITIONAL WAIVER AND RELEASE

This document waives and releases lien, stop payment notice, and payment bond rights the claimant has for labor and service provided, and equipment and material delivered, to the customer on this job through the Through Date of this document. Rights based upon labor or service provided, or equipment or material delivered, pursuant to a written change order that has been fully executed by the parties prior to the date that this document is signed by the claimant, are waived and released by this document, unless listed as an Exception below. This document is effective only on the claimant's receipt of payment from the financial institution on which the following check is drawn:

Maker of Check:
Amount of Check: $
Check Payable to:

Exceptions

This document does not affect any of the following:

(1) Retentions.

(2) Extras for which the claimant has not received payment.

(3) The following progress payments for which the claimant has previously given a conditional waiver and release but has not received payment:

Date(s) of waiver and release:

Amount(s) of unpaid progress payment(s): $

(4) Contract rights, including (A) a right based on rescission, abandonment, or breach of contract, and (B) the right to recover compensation for work not compensated by the payment.

Signature:
Claimant's Signature:
Claimant's Title:
Date of Signature:

Appendix C

Use AFTER July 2012

CIVIL CODE SECTION 8134

If the claimant is required to execute a waiver and release in exchange for, or in order to induce payment of, a progress payment and the claimant asserts in the waiver that the claimant has, in fact, been paid the progress payment, the waiver and release shall be null, void, and unenforceable unless it is in substantially the following form, with the text of the "Notice to Claimant" in at least as large a type as the largest type otherwise in the form.

UNCONDITIONAL WAIVER AND RELEASE ON PROGRESS PAYMENT

NOTICE TO CLAIMANT: THIS DOCUMENT WAIVES AND RELEASES LIEN, STOP PAYMENT NOTICE, AND PAYMENT BOND RIGHTS UNCONDITIONALLY AND STATES THAT YOU HAVE BEEN PAID FOR GIVING UP THOSE RIGHTS. THIS DOCUMENT IS ENFORCEABLE AGAINST YOU IF YOU SIGN IT, EVEN IF YOU HAVE NOT BEEN PAID. IF YOU HAVE NOT BEEN PAID, USE A CONDITIONAL WAIVER AND RELEASE FORM.

Identifying Information
Name of Claimant:
Name of Customer:
Job Location:
Owner:
Through Date:

UNCONDITIONAL WAIVER AND RELEASE

This document waives and releases lien, stop payment notice, and payment bond rights the claimant has for labor and service provided, and equipment and material delivered, to the customer on this job through the Through Date of this document. Rights based upon labor or service provided, or equipment or material delivered, pursuant to a written change order that has been fully executed by the parties prior to the date that this document is signed by the claimant, are waived and released by this document, unless listed as an Exception below. The claimant has received the following progress payment: $

EXCEPTIONS

This document does not affect any of the following:

(1) Retentions.

(2) Extras for which the claimant has not received payment.

(3) Contract rights, including (A) a right based on rescission, abandonment, or breach of contract, and (B) the right to recover compensation for work not compensated by the payment.

Signature:

Claimant's Signature:

Claimant's Title:

Date of Signature:

Appendix D

Use AFTER July 2012

Civil Code Section 8136

If the claimant is required to execute a waiver and release in exchange for, or in order to induce payment of, a final payment and the claimant is not, in fact, paid in exchange for the waiver and release or a single payee check or joint payee check is given in exchange for the waiver and release, the waiver and release shall be null, void, and unenforceable unless it is in substantially the following form.]

CONDITIONAL WAIVER AND RELEASE ON FINAL PAYMENT

NOTICE: THIS DOCUMENT WAIVES THE CLAIMANT'S LIEN, STOP PAYMENT NOTICE, AND PAYMENT BOND RIGHTS EFFECTIVE ON RECEIPT OF PAYMENT. A PERSON SHOULD NOT RELY ON THIS DOCUMENT UNLESS SATISFIED THAT THE CLAIMANT HAS RECEIVED PAYMENT.

Identifying Information

Name of Claimant:

Name of Customer:

Job Location:

Owner:

CONDITIONAL WAIVER AND RELEASE

This document waives and releases lien, stop payment notice, and payment bond rights the claimant has for labor and service provided, and equipment and material delivered, to the customer on this job. Rights based upon labor or service provided, or equipment or material delivered, pursuant to a written change order that has been fully executed by the parties prior to the date that this document is signed by the claimant, are waived and released by this document, unless listed as an Exception below. This document is effective only on the claimant's receipt of payment from the financial institution on which the following check is drawn:

Maker of Check:

Amount of Check: $

Check Payable to:

Exceptions

This document does not affect any of the following:

Disputed claims for extras in the amount of: $

Signature

Claimant's Signature:

Claimant's Title:

Date of Signature:

Appendix E

Use AFTER July 2012

Civil Code Section 8138

If the claimant is required to execute a waiver and release in exchange for, or in order to induce payment of, a final payment and the claimant asserts in the waiver that the claimant has, in fact, been paid the final payment, the waiver and release shall be null, void, and unenforceable unless it is in substantially the following form, with the text of the "Notice to Claimant" in at least as large a type as the largest type otherwise in the form.

UNCONDITIONAL WAIVER AND RELEASE ON FINAL PAYMENT

NOTICE TO CLAIMANT: THIS DOCUMENT WAIVES AND RELEASES LIEN, STOP PAYMENT NOTICE, AND PAYMENT BOND RIGHTS UNCONDITIONALLY AND STATES THAT YOU HAVE BEEN PAID FOR GIVING UP THOSE RIGHTS. THIS DOCUMENT IS ENFORCEABLE AGAINST YOU IF YOU SIGN IT, EVEN IF YOU HAVE NOT BEEN PAID. IF YOU HAVE NOT BEEN PAID, USE A CONDITIONAL WAIVER AND RELEASE FORM.

Identifying Information

Name of Claimant:

Name of Customer:

Job Location:

Owner:

UNCONDITIONAL WAIVER AND RELEASE

This document waives and releases lien, stop payment notice, and payment bond rights the claimant has for all labor and service provided, and equipment and material delivered, to the customer on this job. Rights based upon labor or service provided, or equipment or material delivered, pursuant to a written change order that has been fully executed by the parties prior to the date that this document is signed by the claimant, are waived and released by this document, unless listed as an Exception below. The claimant has been paid in full.

This document does not affect the following:

Disputed claims for extras in the amount of: $

Signature:

Claimant's Signature:

Claimant's Title:

Date of Signature:

Appendix F

Mechanic Lien & Stop Notice Timeline

Two Prerequisites:

- Properly licensed
- 20-Day preliminary notice served

When the Owner Records a Notice of Completion Or a Notice of Cessation

For General Contractors

- Owner files Notice of Completion or Notice of Cessation
- You have 60 days to Record the Mechanic Lien
- Before you Record the Lien, you must provide NOTICE to the Owner. [Any Mechanic Lien is invalid without it!!!]
- Record the Mechanic Lien in the Recorders or Assessor's Office of the County or Town where the project is located.
- File Your Lawsuit within 90 days of recording the Mechanics Lien
- Record a Lis Pendens in the Recorders or Assessor's Office

For Subcontractors & Suppliers

The same sequence of events needs to occur ***EXCEPT*** you have only 30 Days to Record the Lien following Notice of Completion or Notice of Cessation.

When owner accepts project or takes beneficial occupancy without filing a notice of completion or cessation

For General, Subcontractors & Suppliers

- Owner does not file Notice of Completion or Notice of Cessation
- You have 90 days to Record the Mechanic Lien
- Before you Record the Lien, you must provide NOTICE to the Owner.
- Record the Mechanic Lien in the Recorders or Assessor's Office of the County or Town where the project is located.
- File Your Lawsuit within 90 days of recording the Mechanics Lien
- Record a Lis Pendens in the Recorders or Assessor's Office

When the owner does not file notice of completion or cessation, does not take possession and work stops for 60 days

For General, Subcontractors & Suppliers

- Work stops for a continuous 60 days ='s completion
- All contractors have 90 days to Record Lien from end of 60 day cessation (150 day total- 60 + 90)
- Before you Record the Lien, you *must* provide NOTICE to the Owner.
- Record the Mechanic Lien in the Recorders or Assessor's Office of the County or Town where the project is located.
- File Your Lawsuit within 90 days of recording the Mechanics Lien
- Record a Lis Pendens in the Recorders or Assessor's Office

About the Author

John P. McGill is a journeyman carpenter, licensed general contractor, attorney, and presently is outside general counsel for a large public works general contractor and represents a number of subcontractors and general contractors. He received his JD with distinction from the University of the Pacific, McGeorge School of Law, and is a member of the school's Traynor Honor Society. His practice is, and has been, almost exclusively in representing contractors, general and subcontractor, in both private and public work disputes as well as in employment, transactional, and construction-related administrative actions.

McGill has written a number of articles on construction law issues and on appellate decisions related to the construction industry. He has been involved as an attorney in several recent appellate cases on important issues related to the construction industry. He has taught construction scheduling, construction estimating, and construction law at local community colleges in the North Bay Area and continues to provide numerous seminars and presentations on a number of issues related to the construction industry.

McGill lives in Napa California with his wife and two daughters.

Index